Science-Report aus dem Faserinstitut Bremen

Band 15

Science-Report aus
dem Faserinstitut Bremen

Band 15

Herausgegeben von
Prof. Dr.-Ing. Axel S. Herrmann

Holger Büttemeyer

Impregnation of stitched continuous carbon fibre textiles by Sheet Moulding Compounds

Logos Verlag Berlin

Bibliografische Information der Deutschen Nationalbibliothek

Die Deutsche Nationalbibliothek verzeichnet diese Publikation in der Deutschen Nationalbibliografie; detaillierte bibliografische Daten sind im Internet über http://dnb.d-nb.de abrufbar.

ISBN 978-3-8325-5354-8
ISSN 1611-3861

Logos Verlag Berlin GmbH
Georg-Knorr-Str. 4, Geb. 10, 12681 Berlin

Tel.: +49 (0)30 / 42 85 10 90
Fax: +49 (0)30 / 42 85 10 92
https://www.logos-verlag.de

Impregnation of stitched continuous carbon fibre textiles by Sheet Moulding Compounds

Vom Fachbereich Produktionstechnik

der

UNIVERSITÄT BREMEN

zur Erlangung des Grades
Doktor-Ingenieur
genehmigte

Dissertation

von Holger Büttemeyer, M.Sc.

Erstgutachter: Univ.-Prof. Dr.-Ing. Axel S. Herrmann, Universität Bremen
Zweitgutachter: Univ.-Prof. Dr.-Ing. Jürgen Fleischer, Karlsruher Institut für Technologie (KIT)

Tag der mündlichen Prüfung: 12.04.2021

Für

Charlotte und Dorothee

Danksagung

Diese Dissertation entstand während meiner Tätigkeit als wissenschaftlicher Mitarbeiter am Faserinstitut Bremen zwischen 2016 und 2020.

Mein besonderer Dank gilt Herrn Prof. Dr.-Ing. Herrmann für die fachliche Betreuung und den zahlreichen Diskussionen mit ihm. Seine Anregungen und Erfahrungswerte trugen maßgeblich zum Gelingen dieser Arbeit bei.

Für die freundliche Bereitschaft zur Übernahme des Korreferates danke ich Herrn Prof. Dr.-Ing. Fleischer.

Des Weiteren danke ich herzlich meinen Kolleginnen und Kollegen am Faserinstitut für Ihre Unterstützung. Für die zahlreichen und zeitintensiven Laboranalysen danke ich Petra Hartwig, Manuela von Salzen und Frank Rosenkötter. Ich danke Oliver Focke für seine Unterstützung bei der Porenanalyse durch die Röntgenmikroskopie, Dr.-Ing. Adli Dimassi für die kompetente Unterstützung bei der Struktursimulation und Dr.-Ing. Michael Koerdt für seine Mitarbeit bei den experimentellen Untersuchungen zur dielektrischer Analyse. Weiterhin danke ich Dr.-Ing. Patrick Schiebel, Katharina Arnaut, Anna Lang, Robert Gaitzsch, Richard Vocke, Lars Bostan, Tim Frerich sowie Nadine Gushurst, dass sie mir kontinuierlich als Diskussionspartner zur Seite gestanden haben. Vor allem die Gespräche zwischen Tür und Angel brachten diese Arbeit voran.

Für das großartige Projektkonsortium und die tolle Zusammenarbeit innerhalb des Forschungsprojektes EFFKAB, gefördert vom BMWi, welches die inhaltliche Basis meiner Dissertation darstellte, möchte ich mich an dieser Stelle bedanken. Namentlich sind hier Marc Fette (CTC GmbH), Dr.-Ing. Daniel Krause, Dr.-Ing. Tobias Wille (beide DLR Braunschweig) und Gunnar Fick (3D ICOM GmbH & Co. KG) zu nennen. Für die durchgehende Unterstützung bedanke ich mich bei der Polynt Composites Germany GmbH, insbesondere bei Nicole Stöß, Daniel Bröder und Arno Walther.

Herzlichst bedanke ich mich bei meiner Familie und meinen Freunden, besonders meinen Eltern, welche mir meinen beschrittenen Bildungsweg ermöglicht haben. Ihr Vertrauen hat mich bei der Verwirklichung meiner beruflichen Ziele immer unterstützt. Bei Dr. Fokko Eller bedanke ich mich für die zeitintensive Prüfung des Gesamtdokuments.

Abschließend möchte ich mich bei meiner Partnerin Nadine bedanken, die mich gerade in der Endphase dieser Arbeit durch alle Höhen und Tiefen begleitet hast. Neben deinen fachlichen Hinweisen hast du mir mit deiner positiven Art immer Mut zugesprochen und dadurch sehr zur Fertigstellung der Dissertation beigetragen.

Kurzfassung

In der Dissertation wird die Imprägnierung eines Verstärkungseinlegers aus Kohlenstofffasern durch ein Sheet Moulding Compound (SMC) untersucht. Dabei besitzt der Kohlenstoffstruktureinleger zu Beginn keine Vorimprägnierung. Stattdessen erfolgt die Durchtränkung der Kohlenstofffasern durch die Harzbestandteile des SMC-Materials während des Fließpressprozesses. Durch die Kombination des SMC-Materials mit einem Kohlenstofffaserstruktureinleger, welcher mit dem Tailored Fibre Placement-Verfahren hergestellt wird, entsteht ein sogenannter Hybrid SMC Verbundwerkstoff, welcher sich durch eine hohe Designfreiheit, gute mechanische Eigenschaften und gleichzeitig hohen Produktionsraten auszeichnet.
Das Hauptziel dieser Dissertation ist die Entwicklung eines Imprägnierungsmodells für Hybrid SMC Verbundwerkstoffe. Dabei sagt das Imprägnierungsmodell unter Berücksichtigung der Halbzeug- und der Prozesseigenschaften den verbleibenden Porengehalt voraus. Die Imprägnierung des Struktureinlegers ist von der Viskosität, des Prozessdrucks, der Permeabilität und der Dicke des Struktureinlegers abhängig. Dabei ist die Viskosität ein entscheidender Faktor für die Faserimprägnierung, da sie sowohl von der Temperatur als auch von der Zeit abhängig ist. Zum einen bestimmt die Temperatur den Gelpunkt, da nach Überschreiten dieses Punktes keine weitere Durchtränkung der Kohlenstofffasern möglich ist. Zum anderen wirkt sich die Zeit- und Temperaturabhängigkeit auf das generelle Fließverhalten aus, welches üblicherweise für SMC-Materialien hochviskos ist. Thermische Analysemethoden werden daher umfassend eingesetzt, um den Gelpunkt und die Fließeigenschaften zu bestimmen. Auch die Bestimmung der weiteren Parameter fließt in das Imprägnierungsmodell ein. Auf Basis eines Ansatzes aus der Fluiddynamik wird das Imprägnierungsmodell entwickelt, um die Fließfront innerhalb des SMC-Materials während des Fließpressens zu verfolgen. Parallel werden experimentelle Untersuchungen an Hybrid SMC Verbundwerkstoffen umgesetzt und der Porengehalt durch den Einsatz von mikroskopischen, bildgebenden Methoden bestimmt. Die Auswertung erfolgt durch die Anwendung Methoden aus der Statistik, um die signifikanten Parameter zu identifizieren. Zudem wurden die experimentell ermittelten Porengehälter zur Validierung des Imprägnierungsmodells eingesetzt. Insgesamt führten die Untersuchungen zu einem Imprägnierungsmodell mit hoher Genauigkeit. Bei über 82 % aller Messungen stimmt das analytische Imprägnierungsmodell mit einer hohen Genauigkeit (Die Abweichungen sind geringer als 5 %) überein.

Abstract

This thesis deals with the fibre impregnation of a carbon fibre reinforcement by a sheet moulding compound (SMC). In the beginning, the carbon fibre reinforcement has no impregnation. Instead, the impregnation of the carbon fibre is performed by the resin within the SMC material during compression moulding. The combination of a SMC material and a carbon fibre reinforcement, which is made by the Tailored Fibre Placement technology, leads to a Hybrid SMC composite, which is characterized by a high design freedom, good mechanical properties, and high production rates at the same time.
The main objective of this study is the development of an analytical impregnation model for Hybrid SMC composites. The impregnation model predicts the final void content with regard to the properties of the semi-finished products and the process implementation. The fibre impregnation is influenced by the viscosity of the SMC material, the processing compression, the permeability, and the thickness of the carbon fibre reinforcement. Among all these parameters, the viscosity is an essential factor for the fibre impregnation, because it is dependent on the temperature and the time. First, temperature has an impact on the gelation point. After passing the gelation point, no more fibre impregnation can be performed. Second, the time- and temperature-dependency acts on the general flow characteristics which is typically high-viscous for SMC materials. Therefore, thermal analysis is comprehensively used to determine the gelation point as well as the flow characteristics of the SMC material. The other influencing parameters are determined to support the fibre impregnation model. An impregnation model is developed by a fluid dynamic's approach to track the flow front particles within the SMC material during compression moulding. At the same time, experiments are realized and the void content is determined by using microscopic analysis of the Hybrid SMC composites. The experiments are evaluated by the consequent use of statistical instruments to find the most significant parameters. Furthermore, the evaluated void contents of the experiments are used as a validation for the impregnation model. All in all, the investigations have led to an analytical impregnation model with a high accuracy. A deviation of 5 % for more than 82 % of the specimens was achieved.

Table of Contents

List of abbreviations

AFP	Automated fibre placement
ANOVA	Analysis of variance
ATH	Aluminium hydroxide (Alumina Trihydrate)
ATL	Automated tape laying
BMC	Bulk moulding compound
CCFR	Continuous carbon fibre reinforcement
DEA	Dielectric analysis
DSC	Differential scanning calorimetry
CAGR	Compound annual growth rate
CF	Carbon fibre
CFRP	Carbon fibre reinforced plastic
CoDiCoFRP	Continuous-discontinuous long fibre reinforced polymer structures
DoE	Design of experiment
DoF	Degree of freedom
EDX	Energy dispersive X-ray spectroscopy
FRP	Fibre reinforced plastics
FST	Flame, smoke density, and toxicity
FVC	Fibre volume content
GF	Glass fibre
GFRP	Glass fibre reinforced plastic
ICRC	International Composites Research Centre
IMC	In-mould coating
IRTG	International Research Training Group
KIT	Karlsruhe Institute of Technology
LPA	Low-Profile Additives
MEKP	Methyl ethyl ketone peroxide

MQ	Sum of mean squares
OEM	Original equipment manufacturer
OHSC	Overhead stowage compartment
PD	Production direction
RTI	Resin transfer impregnation
RTM	Resin transfer moulding
SD	Standard deviation
SMC	Sheet Moulding Compound
SQ	Sum of squares
TFP	Tailored fibre placement
TGA	Thermogravimetric analysis
XRM	X-Ray Microscope

List of notations

Symbol	Unit	Description
A_{CM}	Pa s	Empirical parameter of the Castro-Macosko-model
A_{SMC}	mm^2	Cross section of the semi-finished SMC material
$A_{SMC,new}$	mm^2	Cross section of the SMC cover layer in Hybrid SMC composites
A_1	1/s	Empirical parameter of the 1st reaction of the expanded Prout-Tompkins-Equation
A_2	1/s	Empirical parameter of the 2nd reaction of the expanded Prout-Tompkins-Equation
a	m/s^2	Acceleration
C_1	1	Empirical parameter of the 1st reaction of the expanded Prout-Tompkins-Equation
C_2	1	Empirical parameter of the 2nd reaction of the expanded Prout-Tompkins-Equation
c	1	Total amount of specimens
CM_1	1	Empirical parameter of the Castro-Macosko-model
CM_2	1	Empirical parameter of the Castro-Macosko-model
d_{fb}	mm	Fibre bundle distance
d_{sl}	mm	Stitch length
E	MPa	Tensile modulus
E_1	J/mol	Empirical parameter of the 1st reaction of the expanded Prout-Tompkins-Equation
E_2	J/mol	Empirical parameter of the 2nd reaction of the expanded Prout-Tompkins-Equation
e	1	Euler's number
F	1	Filler properties
F_1, F_2, F_3	MPa	Fibre prestress

f	1	Function
G'	Pa s	Storage modulus
G''	Pa s	Loss modulus
H	J/g	Heat flow
H_R	J/g	Enthalpy of reaction
H_{total}	J/g	Total heat flow
K	m^2	Permeability
K_z	m^2	Permeability through the thickness of the reinforcement
k_1	1	Empirical parameter of the 1st reaction of the expanded Prout-Tompkins-Equation
k_2	1	Empirical parameter of the 2nd reaction of the expanded Prout-Tompkins-Equation
m	1	Empirical parameter of the time- and temperature-dependent viscosity
m_1	1	Empirical parameter of the 1st reaction of the expanded Prout-Tompkins-Equation
m_2	1	Empirical parameter of the 2nd reaction of the expanded Prout-Tompkins-Equation
n	1	Empirical parameter of the time- and temperature-dependent viscosity
n_L	1	Amount of carbon fibre layers
n_1	1	Empirical parameter of the 1st reaction of the expanded Prout-Tompkins-Equation
n_2	1	Empirical parameter of the 2nd reaction of the expanded Prout-Tompkins-Equation
p / P	Pa	Processing compression
p_{app}	Pa	Applied procession compression
p_{fluid}	Pa/m	Pressure gradient of the fluid
Q_{SMC}	g/m^2	Surface weight
R	J/(mol K)	Ideal gas constant
R_m^2	1	Coefficient of determination for the empirical parameter m
R_n^2	1	Coefficient of determination for the empirical parameter n
R_p^2	1	Coefficient of determination for regression model
r_{cf}	m	Radius of the carbon fibre's cross section
T	°C	Temperature
T_b	K	Empirical parameter of the Castro-Macosko-model
T_{gel}	°C	Temperature at the gelation point

T_K	K	Temperature
t	s	Time
t_{gel}	s	Time at the gelation point
V	m/s	Flow rate or fluid's velocity
v_p	m/s	Velocity of particles within the flow front during compression moulding
w_{cf}	mm	Width of a carbon fibre bundle
y_i	1	Result of the experimental measurements i
$\bar{y}$	1	Mean results of the experimental measurements
z	mm	Impregnation length
z_{bm}	mm	Thickness of the base material
z_{CF}	mm	Thickness of the impregnated carbon fibre bundles
z_{ff}	mm	Position of particles within the flow front during compression moulding
z_{half}	mm	Half of the total thickness of the impregnated fibre reinforcement which contains carbon fibre bundles and base material
z_{SMC}	mm	Thickness of the semi-finished SMC material
$z_{SMC,new}$	mm	Thickness of the SMC cover layer in Hybrid SMC composites
z_{total}	mm	Total thickness of the impregnated fibre reinforcement which contains carbon fibre bundles and base material

Symbol	Unit	Description
α	%	Degree of cure
α_{gel}	%	Degree of cure at the gelation point
α_{sl}	%	Significance level
β	1	Emprical parameter of the Kästner-Equation
$\dot{\gamma}$	1/s	Shear rate
Δ	1	Describes a difference of two values
δ	1	Loss angle
ε	%	Strain
η	Pa s	Viscosity
η_c	Pa s	Viscosity considering cure effects
η_f	Pa	Viscosity considering filler effects
η_{ion}	Ohm cm	Ionic viscosity
η_{sr}	Pa s	Viscosity considering shear rate effects
η_0	Pa s	Zero viscosity
ϑ	1/mm^2	Stiching density

κ	%	Void Content
λ	1	Ideal Box-Cox-Exponent
ξ	1	Empirical parameter of the developed fibre impregnation model
ρ_F	g/cm^3	Fibre density
ρ_{SMC}	g/cm^3	Density of the SMC material
σ	MPa	Tensile Strength
φ_{CF}	1	Carbon fibre volume content
φ_{GF}	1	Glass fibre volume content
$\varphi_{GF,new}$	1	Glass fibre content after compression moulding
ψ_F	1	Fibre weight content

Chemical symbol	**Description**
C	Carbon
CH	Methylidyne, or (unsubstituted) carbyne
CH_2	Methylene
H	Hydrogen
H_2O	Water
MgO	Magnesium oxide
O	Oxygene
OH	Hydroxide
R	Rest, which is not influenced by the main reaction

1 Introduction

1.1 Motivation

Productivity is one of the most challenging factors in the development of fibre reinforced plastics (FRP). It is defined as the ratio between the output and the input of a good [1]. Taking FRP into account, the output can be frequently found in the low part weight and the excellent mechanical properties at the same time. However, the dimensioning of FRP is more complex due to the anisotropic fibre properties. In addition, especially carbon fibres are more expensive due to their extensive production process. Therefore, the use of FRP is often hindered. This is why it is important to find new ways to increase FRP productivity and to find fields in which the use of FRP can have a significant impact.
Hybrid solutions are an appropriate method to increase the FRP productivity. A hybridization is the combination of at least two materials or processes to use the advantages of both [2]. In the ideal case, the specific disadvantages of the individual materials are eliminated due to the combination. For example, the development of intrinsic hybrids is preferred which are characterized by one-step processes which do not require any subsequent joining processes [3]. In 1984, Tompkins and White has developed the design rule "The right material at the right place" which links to the approach of hybridization [4]. This approach has been asserted oneself in the engineering environment, especially in the automotive industry, which demonstrates the success of the hybrid method of construction [5]–[7].
The manufacturing of Hybrid Sheet Moulding Compound (Hybrid SMC) composites is a promising technology to realize powerful composites with an economic production process at the same time. It synergises a SMC material and

a continuous fibre reinforcement. In this case, the continuous carbon fibre reinforcement is manufactured with the Tailored Fibre Placement (TFP) technology. This technology allows the production of the near-net shape and load-path optimized designs. Therefore, the production is characterized by a good material efficiency. Due to the carbon fibre reinforcement, the mechanical properties can be significantly increased in comparison to pure SMC materials [8]. Moreover, short cycle times for the manufacturing of SMC parts can still be achieved. From an ecologic perspective, the compression moulding of SMC materials is an energy-efficient production process in comparison to other FRP manufacturing technologies [9]. Metallic inserts can be easily implemented to allow a simple joining for other components. This brief description of the advantages illustrates the great potential of Hybrid SMC composites.
However, new hybrid material solutions require an accurate safeguarding of the composite quality. In the case of the manufacturing of Hybrid SMC composites containing a carbon fibre reinforcement, one of the most important quality characteristic is the impregnation of the reinforcing fibres. In general, the fibre impregnation of carbon fibre textiles within Hybrid SMC composites is a challenging task which strongly differs to the fibre impregnation of other thermoset processing. Especially the high viscosity and the low permeability due to the use of a carbon fibre reinforcement as well as the expected high fibre volume content hinder the fibre impregnation. An incomplete fibre impregnation would lead to a decreased fibre support which finally results in reduced mechanical properties [10]–[12]. Therefore, it is important to ensure the impregnation of the reinforcing fibres to avoid the formation of voids and to avoid premature failure of the hybrid composite.

1.2 Objective

The objective of this study is the development of an analytical model to predict the fibre impregnation of dry continuous fibre reinforcements by using conventional SMC materials as resin carriers. It considers the important material and process characteristics which significantly influence the fibre impregnation of continuous carbon fibre reinforcements while compression moulding. The analytical model is tested through additional experiments.

The individual results can be summarised as follows:

- Characterisation of Hybrid SMC containing TFP reinforcements and SMC
- Formulation of a methodology for the development of an impregnation model
- Determination of the chemo-rheological behaviour of SMC
- Development of an impregnation model
- Experimental investigation to evaluate the key parameter

2 Fundamentals of Sheet Moulding Compounds

Compression Moulding of Sheet Moulding Compounds is one of the oldest manufacturing techniques for processing fibre reinforced plastics. By 1947, a composite body of a car, mostly based on SMC, has already been developed and tested for the first time which finally led to the Corvette C1 [13]. Today, new fields and applications are implemented through the use of SMC in the automotive industry. Recently, Lamborghini has searched for a lightweight material for the manufacturing of a rear wing of the model Huracan Performante. Besides the traditional structural and cosmetic requirements, the desired material has to be formed to a hollow structure. This hollow structure is used for the active wing system to ensure maximum down force or either minimum drag. The compression moulding of SMC enables the manufacturing of composite structures to fulfil these requirements due to high design freedom [14]. But not only the supercar manufacturers use carbon fibre SMC materials. Recently, Toyota has presented their new method for the manufacturing of a rear door frame made out of carbon fibre SMC [15]. It indicates that the technology even enables large scale carbon fibre components in non-luxury passenger cars (Figure 1, right).
However, just 21 % of the annually produced SMC materials are used in the automotive industry. Further important industries are the electrical/electronic, the aviation/defence, and the construction sector (*Figure 2*) [16]. In Europe, the annual growth rate of SMC is 2.2 % per year. More or less, this growth rate was constant in the last years [17]. In the next ten years, the compound annual growth rate (CAGR) will rise to 6.0 % (Figure 2) [16]. Together with material-related BMC, the total amount of the moulding compounds is 280.000 tons per year. Therefore, it is the most frequently used manufacturing method for

processing glass fibres to FRP [5]. An ideal production rate for using SMC is between 10.000 and 80.000 parts per year which is determined by the short cycle times of 2 – 5 minutes [18].

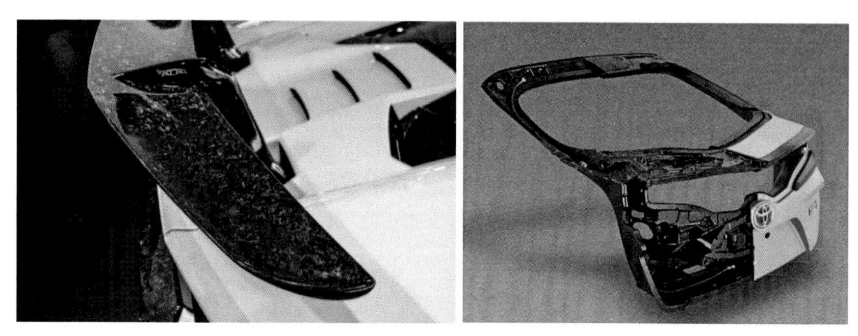

Figure 1. Rear wing of the Lamborghini Huracan Performante (left) [19]; *Rear door frame of the Toyota Prius PVH [15]*

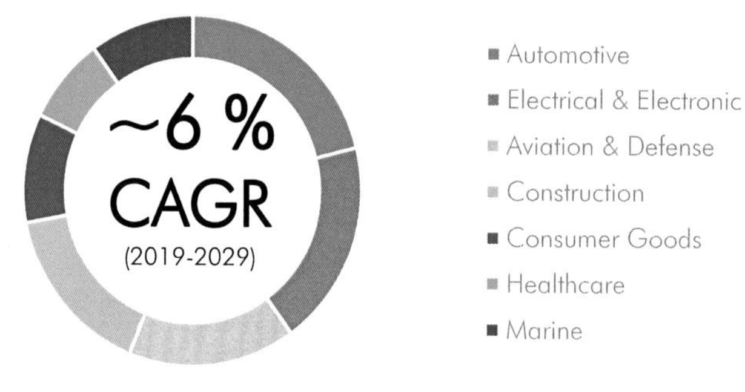

Figure 2. Sheet Moulding Compounds market by end use, 2019 [16]

In the following, the general characteristics of SMC materials are presented. At first, a description of the individual compositions of SMC materials is made. The individual compositions are processed with an impregnation lane for the manufacturing of a semi-finished SMC which is required for the subsequent compression moulding process. In addition, the thickening and curing reactions of unsaturated polyesters are explained because an UP-based SMC material is used in this study. Further information is given about the types of reinforcements. These are predominantly glass fibre reinforcements in the shape of long fibres but other fibres were already used or investigated in the past.

2.1 From Compounding to Material Curing

The denotation "Sheet Moulding Compound" is normed in DIN EN 14598-1:2005-07 which includes the mixture of different materials with their mass contents, the process, and its properties [20]. The different materials can be divided into fibre materials, resin system, additives, and fillers. A brief overview of these materials is given in the following section.

2.1.1 Constituents

Fibres

Sheet Moulding Compounds are semi-finished products with mostly pre-impregnated long fibres by a thermoset resin. Predominantly, glass fibres are used because of good mechanical properties with adequate market price at the same time. The fibre materials have a length of one or two inch (25.4 mm and 50.8 mm). Carbon fibres and natural fibres are an alternative fibre materials instead of glass fibres as well. Carbon fibres have excellent mechanical properties but they are more cost-intensive. Flax fibres partially have similar specific mechanical properties in comparison to glass fibres but the material consistency is still a problem for mechanical loaded applications [21].

Resin

Similar to the fibre materials, several thermosetting resin systems are available for the use in SMC materials. This includes epoxy, vinyl ester, and unsaturated polyester resins. Epoxy resins have a good surface quality and the best mechanical properties. Unsaturated polyesters (UP) are mostly used because of the sufficient mechanical properties and a lower market price [22]. In this study, a SMC material based on an UP-resin has been used that is why UP-based resins are comprehensively described in the next chapters.

Fillers

Usually, filler materials consist of the greatest mass percentage in SMC materials. Due to the addition of fillers, material properties can be significantly influenced. There are many different types of filler materials that is why it follows a short description of the most common filler types. An essential filler material is magnesium oxide (sometimes zinc oxide as well). Magnesium oxide reacts with unsaturated polyester which leads to the start of the thickening reaction right after the production of the semi-finished SMC [23]. The thickening reaction results in a strong increase of the viscosity. After the thickening reaction is completed, the SMC material is manageable for compression moulding. Calcium carbonate or chalk is often used in SMC materials. Usually, calcium carbonate is considered because of its low material price. Furthermore, it is nontoxic, odourless, has a white colour with a low refractive index, soft, dry, and stable over a wide range of temperatures [24]. Some further effects can be found in the consistent resin flow during compression moulding and higher surface qualities of the finished composite [25]. The addition of hollow glass pearls in the mixture is a good method to decrease the weight of the overall SMC material [26]. Recent research has been done on the development of alternative filler materials which finally led to a reduction of the density and have a finer grain boundary structure at the same time [25]. Aluminium hydroxide can be added to the resin mixture to improve the fire resistance but it negatively affects the viscosity and the weight [27].

Additives

Due to the curing reaction of the UP resin, the SMC material shrinks. Usually, the shrinkage can be about 7.0 – 10.0 % [28]. Therefore, low-profile additives (LPA) are added to the compound to counteract the material shrinkage. That is caused due to a micro void formation between the UP and the LPA phase which counteracts the crosslinking shrinkage. However, these micro voids within the cured composite lead to a reduction of the mechanical performance [29].

The desired part properties can be tailored by using the right constituents with its part amount. Even if the recipes for the production of a semi-finished SMC can strongly differ, Figure 3 shows an approximate composition of a standard SMC material. It catches the eye that inorganic fillers do often have the highest amount because the processing properties become better and fillers have often a low market price. Furthermore, the exemplary SMC recipe from Figure 3 shows the relative low content of the cross-linking resin. However, other compositions for the production of high-performance structural SMC consist of fibre volume contents up to 60 %.

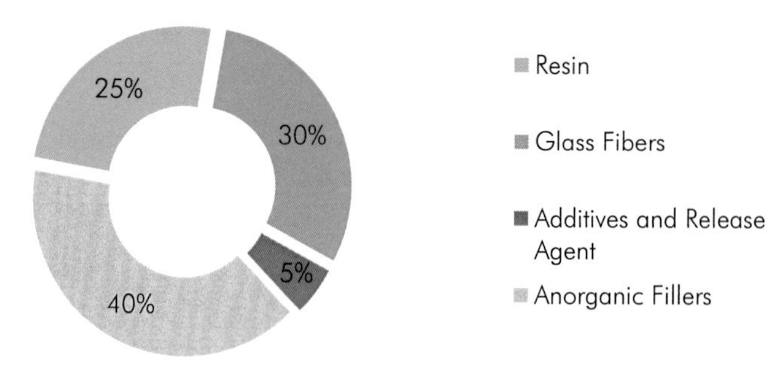

Figure 3. Exemplary composition of a SMC material [30]

2.1.2 Manufacturing of the semi-finished product

Figure 4 shows an impregnation lane for the manufacturing of a semi-finished product of SMC.

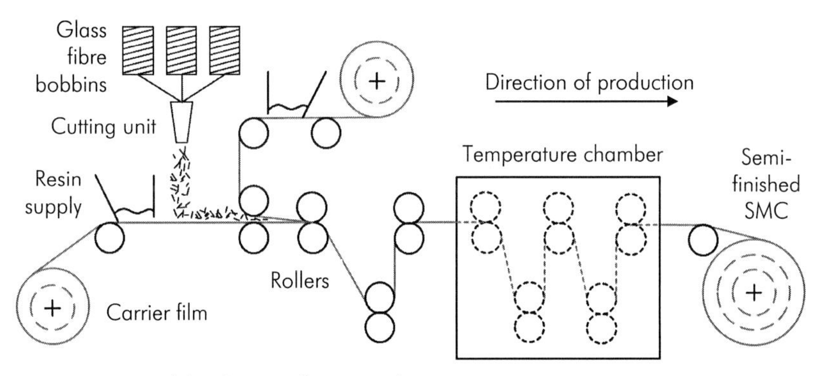

Figure 4. Typical SMC manufacturing lane

The tailored resin mixture is put in boxes behind the carrier films. Both carrier films usually contain polyethylene as the base material because of its good sliding friction properties, high elongations, and low-cost purchase [31]. Sometimes, hybrid carrier films are used with polyethylene and polyamide because polyamide is more tear-resistant and intense odour styrene within the SMC can be embanked.
Glass fibre bundles are routed to cutting units which cut the bundles to a constant length. These long fibres fall on the lower and already impregnated carrier film without any predominant direction which almost results in a quasi-isotropic semi-finished product. In this context, Teodorescu worked on the dependency between the speed of the SMC production line and the final fibre orientation. He defined the fibre contact angle and related it to the reorientation angle [32]. Both carrier films with the glass fibres and the resin mixture are guided together and run through several rolling mills where additional pressure can be applied. Sometimes, the rolling mills are encased in a temperature chamber to influence the viscosity of the resin. Quality assurance on the overall SMC material is a big challenge. Experimental investigations have been done to identify material defects with adapted control units. Air entrapping, fibre distribution, and foil tearing have been observed within the impregnation lane [33]. Such defects can be mostly repaired with process adaptions. At the end of the impregnation lane, the semi-finished SMC is stored on rolls. The thickening reaction starts after some minutes to increase the viscosity of the SMC. After the thickening

reaction is finished, the SMC material is ready for the compression moulding process.

2.1.3 Thickening reaction

The thickening reaction leads to an increase of the material's viscosity. It is triggered by a reaction between the acidic groups at the molecular end of unsaturated polyesters and metal oxide or hydroxides. The most popular metal oxide is magnesium oxide because it is more reactive than magnesium hydroxide, calcium hydroxide, or zinc oxide [34]. Therefore, it results in faster thickening reactions [34]. Typically, the thickening reaction takes between two or three days [35]. Recently, researchers have been working on a reduction of the thickening reaction. A storage time between the semi-finished SMC production and the compression moulding of four minutes has been achieved, so that a direct subsequent processing has already been realized [36]. The increase of the viscosity is influenced by several factors such as the concentration of magnesium oxide and water content but there is still a lack of understanding in the accurate interactions of these factors as well as the overall reaction mechanism [37]. An experimental study has just generally shown that excess percentage of magnesium oxide is necessary for thickening reaction and that an addition of water increases the initial rate of thickening but it reduces the maximum viscosity after thickening [23].
In recent years, three theories have been proposed to describe the chemical thickening mechanism between unsaturated polyesters and metal oxides. A theory by Vansco-Szmercsanyi has described the thickening reaction as a two-stage mechanism between the magnesium oxide and the carboxyl groups of the polyester polymers to form very high molecular weight polymeric salts [38]. Gruskiewicz and Collister have added that this reaction is characterized by a high degree of three dimensional entanglement coupling according to the high amount of carboxyl groups along each polymer molecule [39]. In a study by Sueck, he has described a similar explanation. He has explained that the compounds of carboxyl groups and magnesium are hardly soluble with the unsaturated polyester and form an own nanoscale like a cluster which acts as a multifunctional network and is responsible for the strong increase of viscosity after the thickening process [40]. An alternative theory was formulated by Burns et al. who explained the thickening reaction with the formation of a neutral salt by an interaction of bonding between magnesium ions and two polyester molecules to increase the molecular weight [41]. Rao and Gandhi have presented a theory about the formation of ionic sites with close resemblance to other polymers containing metal ions or ionomers [34], [42]. Rodriguez has made a proposal for the chemical thickening reaction mechanism between unsaturated polyesters and magnesium oxide (Figure 5) [34]:

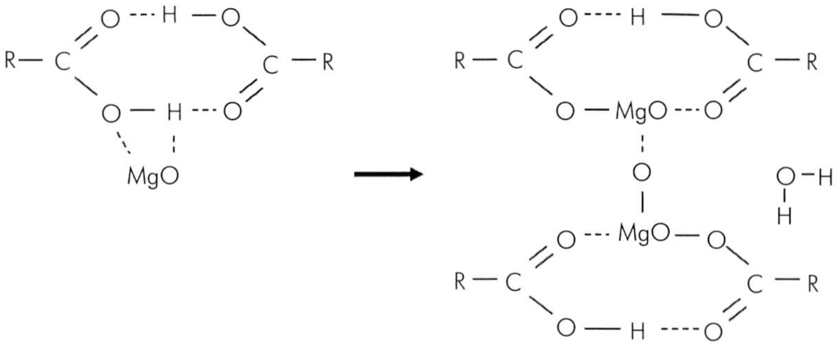

Figure 5. Chemical mechanism for the initial step in the thickening reaction of an unsaturated polyester [34]

2.1.4 Compression Moulding

SMC is processed by compression moulding after the thickening reaction of semi-finished product is completed. The composite manufacturing process can be divided into five specific manufacturing steps. These steps are the cutting of the material, the material insertion into the mould, compression moulding, demoulding, and post-processing of the finished SMC part.

At the beginning of the process, the input of the SMC material has to be calculated which is done by the determination of the SMC weight and the comparison to the final part weight. Then, the cutting of the material is implemented by considering the part weight. Here, automatic cutting units can be used. Usually, the blank size is smaller than the mould cavity to force a material flow. The ratio of the blank size to the total mould cavity is called mould coverage. Usually, the mould coverage is between 60 and 80 % because of two main reasons [43]. On the one hand, the remaining voids from the semi-finished SMC are pushed to the edges of the cavity and finally out of the mould. On the other hand, material flow results in better surface qualities [44]. When mould coverages of almost 100 % is used, higher processing compressions are necessary to reduce the void content [45].

The overall compression moulding process is shown in Figure 6. First, the SMC plies are stacked and subsequently taken to the open mould. The mould is already heated up to the processing temperature and is not changed during the process. The mould is closed and the material flows to the areas of the mould, which are not covered in the beginning of the process. During the material flow, the long fibres realign in the direction of flow.

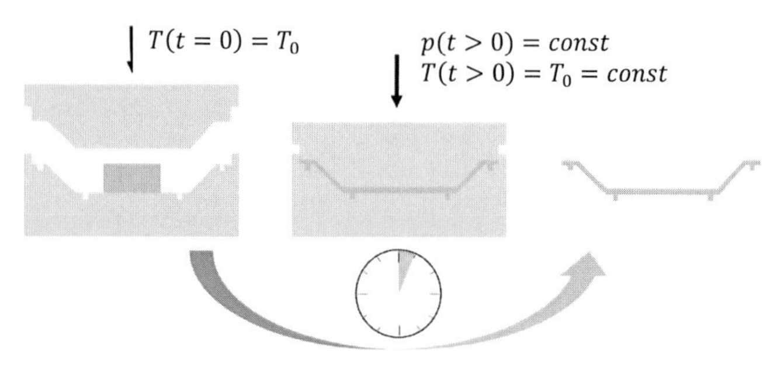

Figure 6. Processing of SMC: Material insertion, compression moulding, demoulding

A recent study has been done by Schladitz et al. who have used a scanning acoustic microscopy and μ-CT. They have combined both analysing techniques with an anisotropic Gaussian filter to determine the fibre orientations in tested specimens [46]. Already in 1987, Kau has developed a computer aided technique for the measurement of fibre orientations in SMC and has determined the strong dependency of flow behaviour and the fibre orientation [47]. The impact of the fibre alignment is certainly big because it significantly influences the mechanical properties. Especially rib designs determine the fibre orientation, when the SMC material flows into it. Curved rib entrances have shown a more arranged fibre orientation than angled ones (Figure 7).

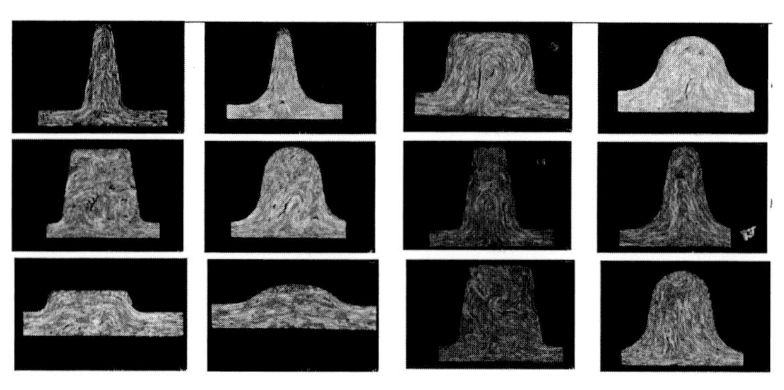

Figure 7. CT images of SMC parts with different rib designs [48]

Evans et al. have found that mould coverages lower than 100 % show higher tensile strength which is caused by higher fibre alignments [45]. Furthermore, many researchers worked on computer-aided engineering tools to predict the final fibre orientations [49], [50]. A rapid mould closing speed is recommended to avoid premature gelation [51]. High processing compressions of 100 bar

are used to avoid void formations. Voids can be found because of several reasons (poor impregnation while the manufacturing of the semi-finished SMC, wetting properties of glass fibres, stacking in the mould, curing mechanism with styrene) [52]. Ferré Sentis et al. have investigated the development of voids in compression moulding processes. They have evaluated two categories of the void development. First, connected voids flow through the open void network to finally expel out of the composite. Second, closed voids decrease the void size and finally coalesced [53]. Furthermore, the type of gas has also an effect on the void content. Carbon dioxide dissolves faster in comparison to air, nitrogen or argon [54].
Compression moulding of SMC is characterized by short cycle times. These cycle times are dependent on the part thickness of the composite [55]. Usually, hydraulic controlled ejectors are used to fulfil the economic demand on a fast demoulding of the cured part and a subsequent preparation for the next curing cycle.
In general, post-processing activities of cured SMC parts are relatively low. Further advantages of SMC parts are the good surface qualities which are certainly dependent on the mould surface [44]. Therefore, further surface treatment is not generally required. In addition, painting of the part can be easily implemented. The design of the mould shearing edges determines the post-processing work as well. Tight tolerances lead to a reduction of material escape. Therefore, a suitable mould shearing edge of 0.02 and 0.05 mm is recommended [56].

2.1.5 Curing reaction

Unsaturated polyester (UP) is the most popular thermoset resin system in SMC materials. To understand the curing mechanisms of the SMC material, it is primarily important to know about the molecular structure of UP resin. In contrast to thermoplastic saturated polyesters, UP-resin are characterized by containing reactive double carbon bonds [57]. UP-resins are gained by a linear polycondensation known as polyesterification reaction which usually requires diols and saturated as well as unsaturated dicarboxylic acids [58]. In general, many types of diols and dicarboxylic acids are available. Figure 8 shows the chemical extraction to gain an UP-resin with terephthalic acid (saturated dicarboxylic acid), maleic acid (unsaturated dicarboxylic acid), and ethylene glycol (diol). The chemical esterification reaction takes place with heat supply and acid catalysis and continuous split-off of water [59]. To cure UP-based resins in SMC materials, a reactive monomer is required. Usually, styrene is used which is an unsaturated hydrocarbon [13]. Therefore, it consists of a reactive double carbon bond (Figure 9).

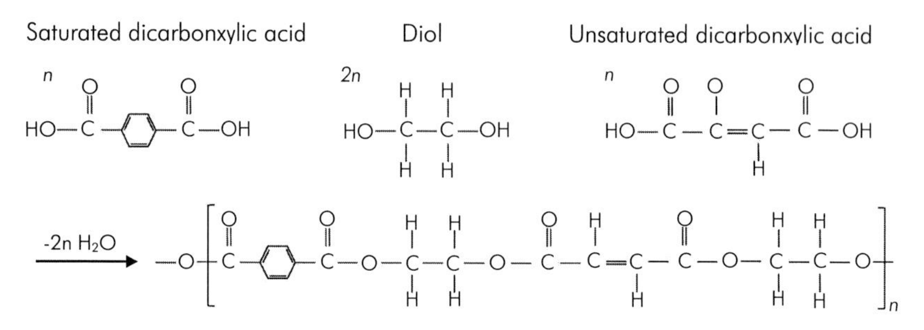

Figure 8. Chemical extraction of unsaturated polyesters [59]

Figure 9. Chemical structure of styrene [60]

The curing mechanism between UP-resin and styrene takes place as a radical polymerisation. A radical polymerisation contains the following four steps [61]:

1. Radical generation from non-radical species (Initiation)
2. Radical addition to a substituted alkene (Propagation)
3. Atom transfer and atom abstraction reaction (Chain transfer and termination by disproportionation)
4. Radical-radical recombination reactions (Termination)

At the beginning of the process, an initiator breaks down due to heat supply and forms radicals. Usually, organic peroxide links are used. Accelerators are used to influence the reaction rate. A typical initiator is methyl ethyl ketone peroxide (MEKP) and cobalt octoate as an accelerator [62]. The general structure of an organic peroxide and its decomposition is shown in Figure 10.

$$R-O-O-R \longrightarrow R-O + O-R$$

Figure 10. General structure of an organic peroxide and its decomposition [63]

The free radicals open the carbon double bond of the styrene monomer to form a new radical which can be linked again with another styrene monomer (Figure 11). This process continues as long as the network encounters an unsaturated polyester chain (Chain-growth reaction) [61]. In the end, a three-dimensional and cross-linked polymer is formed between polystyrene chains and unsaturated polyester chains (Figure 12).

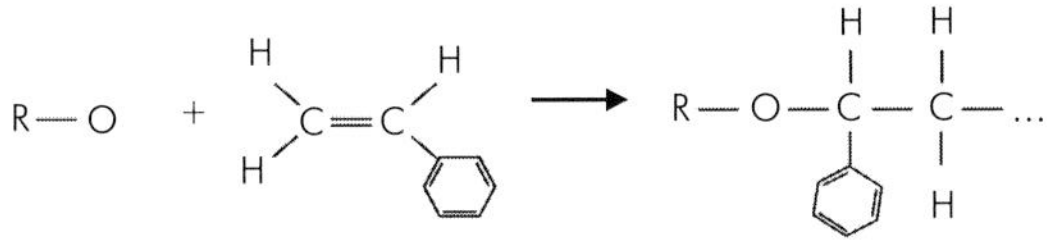

Figure 11. Start of the curing reaction [63]

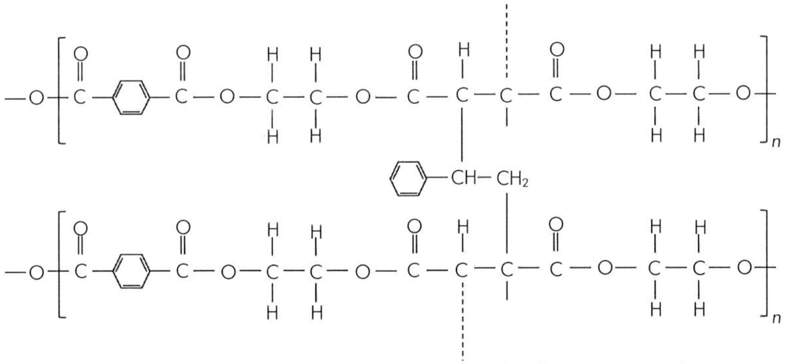

Figure 12. Three-dimensional cross-linked network of UP-resin and styrene [59]

2.2 Types of Reinforcements

Compression moulding of SMC allows many options to design the fibre reinforcement. Both the fibre length and the fibre material can be chosen considering the desired composites properties. This chapter describes the different types of reinforcements. A segmentation of the fibre material as well as the fibre length is done. In the past, classifications were made with respect to the fibre length as well as the degree of fibre orientation (Figure 13) [64].

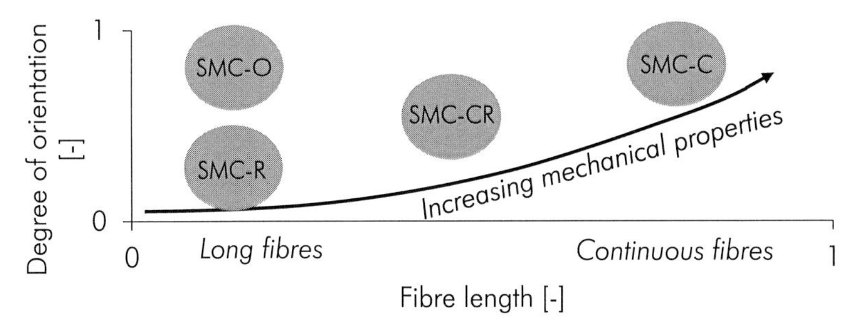

Figure 13. Mechanical properties of different types of SMC [64]

A final chapter considers other types of reinforcements and recent research activities in the field of hybridization of SMC composites.

2.2.1 Long fibre reinforcements

According to DIN EN 14598-1:2005, SMC materials are defined as a reinforced and curable moulding compound in the shape of flat mats [20]. Regarding the fibre length, long fibres are the most common reinforcements. The fibre length significantly decides about the mechanical properties of the SMC composite. Figure 14 demonstrates the dependency of the fibre length and the resulting mechanical properties of stiffness, strength, and impact resistance of glass fibre-polypropylene composite [65]. In contrast to the tensile strength, a high percentage of the overall stiffness level is already achieved at fibre length of 1.0 mm. The typical fibre length of long fibres in SMC materials is either 25.4 or 50.8 mm. Higher fibre lengths do not result in much better mechanical properties because the maximum of the stiffness as well as the strength properties are almost achieved but it would hinder the processing due to compression moulding.

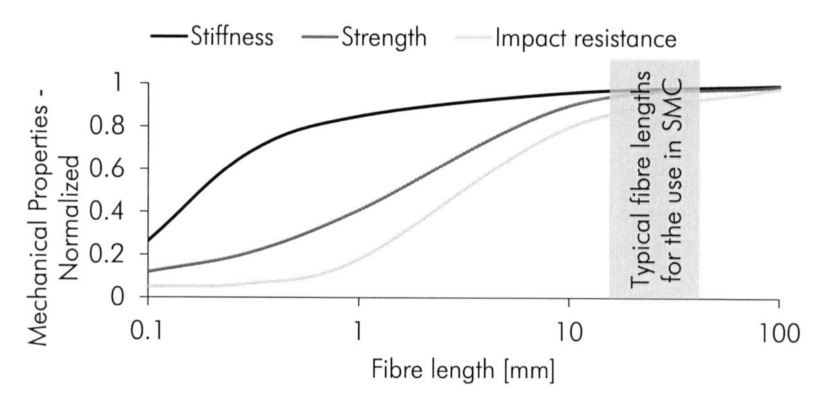

Figure 14. Fibre length significantly decides about the mechanical properties [65]

The fibre volume content is chosen in dependence on the part requirements. The target fibre volume content of standard SMC material is between 20 – 30 %. In some structural cases, the fibre volume content can be raised to 38.5 or 50 weight percentage [66], [67].

Instead of glass fibres, carbon fibres can be used as long fibre reinforcements as well. In comparison to glass fibres, carbon fibres have higher tensile strengths and higher stiffness with a lower density at the same time. But composites with carbon fibres as reinforcements have a 3 to 10 times higher price

than glass fibre reinforced plastics which prohibits the commercial use in many cases [68]. Lee et al. have performed some experimental investigations and have demonstrated that carbon fibres with a length of one inch and a weight content of 50 % achieved a tensile strength of 116.4 MPa, a young's modulus of 12.5 GPa and a flexural strength of 269.4 MPa [69]. Moreover, the fatigue behaviour was investigated considering the degree of fibre orientation by a recent study [70]. In another study, the fracture behaviour of carbon fibre reinforced SMC was studied. It was noticed that the propagation of the crack does not continuously run in comparison to unidirectional composites because of the random orientation of the discontinuous chips [71].
Glass fibre reinforced SMC (GF-SMC) and carbon fibre reinforced SMC (CF-SMC) can be integrated in the same compression moulding process to establish a new hybrid composite type. Two main results were demonstrated in a study of Cabrera-Rios and Castro. First, the combination of GF-SMC and CF-SMC leads to an adaption of the mechanical properties which was significantly influenced by the ratio of carbon fibres to glass fibres. Higher carbon fibre contents led to higher tensile and flexural properties. Furthermore, the ply position of GF-SMC and CF-SMC has a significant effect on the tensile strength. It was demonstrated that the stacking sequence with two CF-SMC plies in the middle results in the highest tensile strengths [72].

Figure 15. Combination of a GF-SMC (core) and a CF-SMC (cover layers) [8]

Moreover, short and long fibres can be a product in the end of the life cycle. The obtained long fibres can be fed in the manufacturing process of the semi-finished SMC product. In a study of Palmer et al., recycled carbon fibres were gained by a mill type granulator. A tailored feeding unit was developed to ensure a uniform feed within the semi-finished SMC product [73], [74]. A commercial production plant for the winning of carbon fibres out of CFRP is the CFK Valley Stade Recycling in Wischhafen. Here, pyrolysis is used to separate the high-cost carbon fibres and the resin. Either the gained carbon fibres can be directly used for the production of a semi-finished SMC material or it can be used to process a non-woven fabric. The non-woven fabric can be used as a reinforcement in the production of a new semi-finished SMC product as well as a textile reinforcement for other composite manufacturing methods. Mechanical tests on processed SMC materials with recycled carbon fibres in the shape of a non-woven fabric have shown that even tensile and flexural properties partially increase in contrast to non-recycled carbon fibre products within SMC [75]. The ratio of price and performance does not meet the industrial requirements at the moment. Further investigations will work on an increase in the mechanical performance, for example due to higher fibre volume contents [76].

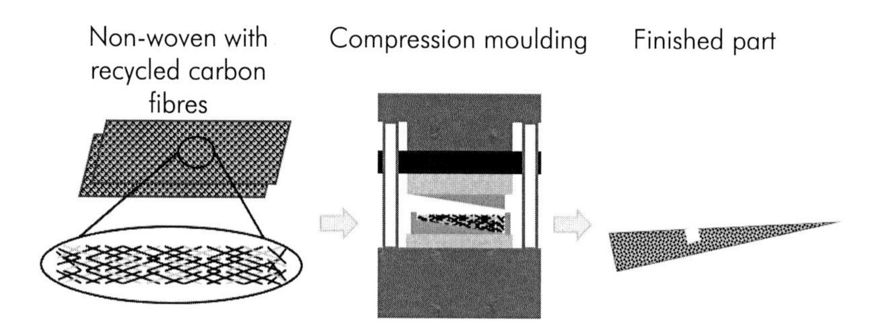

Figure 16. Integration of a non-woven fabric to the compression moulding of a SMC part (According to: [75])

Today, we witness a growing sustainable responsibility in the world. In the future, this will require the investigation of natural fibres as an alternative solution for long fibre reinforcements. Van Voorn et al. have developed a SMC material that uses flax fibres as reinforcing fibre material. Flax fibres have been chosen because the specific tensile and flexural properties are similar to glass fibre materials. Furthermore, flax fibres have a competitive price on the market because of the high availability. In an experimental study by Van Voorn et al., the fibre length, the fibre volume content, and the filler content has been varied and mechanically tested. In comparison to GF-SMC, flax fibre reinforcements in SMC result in an even lower weight (11 – 20 %), however the impact properties are significantly worse [21]. The flame retardancy of natural fibre reinforced SMC was investigated by using hemp fibres. It was proven that the SMC material fulfils all requirements regarding the flame retardancy for the use in building applications [77].
Asadi et al. have made some investigations on basalt fibres as alternative fibre materials within SMC. The investigations have shown similar tensile, flexural, and impact properties in comparison to glass fibres [78]. Also the market price of basalt fibres is similar to that of glass fibres [79]. However, basalt fibres are still a niche within long fibre reinforcements in SMC.

2.2.2 Continuous fibre reinforcements

Due to the fibre length of long fibre reinforced SMC materials, the mechanical properties of the composite are strongly limited. If higher mechanical properties are requested and the advantages of the SMC compression moulding process shall remain, a continuous fibre reinforcement could be a solution. This thesis focuses on the hybridization of a continuous carbon fibre reinforcement with SMC materials. But already in the past, several researchers have worked on this topic because of the great potential for the economic production of high-performance composites. The studies and the results are briefly presented.
The most popular studies were published by an International Research Training Group (IRTG). They worked on a public funded project (Support code: GRK 2078). IRTG is a collaboration of the Karlsruhe Institute of Technology (KIT) in Germany and the International Composites Research Centre (ICRC) in Canada. The project is called "Integrated engineering of continuous-discontinuous long fibre reinforced polymer structures". The main objective of the project was the manufacturing, modelling, and dimensioning of continuous-discontinuous long fibre reinforced polymer structures (CoDiCoFRP). The results were summarized in a scientific book [80] but several findings are described in the following.
The approach, which was used by the IRTG, differs to the approach in this thesis in some aspects (Table 1). First, they have used carbon fibre non-crimp fabrics as reinforcements for specific areas in the compression moulding process. Second, the resin system has been adapted and changed. At least, they have introduced a further production step which was used for the pre-impregnation of the non-crimp fabric. Therefore, the fibre reinforcement was already impregnated before the compression moulding was implemented [80].

Table 1. Comparison of the studies by IRTG and the presented study

	IRTG	This study
SMC material	Self-made	Commercially available
Type of reinforcements	Non-crimp fabrics	TFP fabrics
Pre-impregnation of the reinforcements	Yes	No

Bücheler has modified the resin of the pre-impregnated fibre bundles with ferromagnetic particles to guarantee the position of the reinforcement by magnetic fields during compression moulding [81]. Otherwise, displacements and deformations would occur due to the flow of the SMC during compression moulding [81]. The failure behaviour of CoDiCoFRP has comprehensively been described with specific test equipment. The acoustic emission analysis was

found to be an appropriate method to describe the complex failure mechanism during the three-point bending test. The carbon fibre reinforcements have influenced the damage mechanism due to acoustic recorded signals. These signals were implemented into machine learning algorithms [82]. Further investigations have been done to evaluate the ideal cruciform specimen design for a biaxial tensile tests with high strain levels in the centre region [83]. Another focus of the IRTG was the implementation and evaluation of several techniques of quality assurance during the manufacturing of the semi-finished SMC. Due to the use of thermography and ultrasound, they were able to find out delaminations, inclusions, porosity, and fibre distribution. The study shows that the use of thermography works well to identify fibre distribution and fibre orientation. The best results for the evaluation of delaminations, inclusions, and porosity have been achieved with ultrasonic-testing with air coupling. This ultrasonic testing was superior in comparison to immersion or spectroscopy [84]. A comprehensive study for the determination of the mechanical properties for CoDiCoFRP has been done as well. Here, most of the important mechanical properties have been increased [85]. Moreover, investigations about the machinability have been realized for CoDiCoFRP. Here, the impact of the several machinability parameters has been quantified [86]. Beside the experimental tests and material as well as process characterization, numerical developments have been done by the IRTG as well. A novel simulation approach was developed by Hohberg. It considers the resin-rich lubrication layer with its shear gradient as well as the core region with its plug-flow behaviour [87]. In addition, the local and continuous fibre reinforcements have been considered in the simulation approach [87].
Wulfsberg et al. have investigated a Hybrid SMC with woven fabrics as well as unidirectional carbon fibre reinforcements. As SMC material, they have used HUP27 which is similar to SMC material in this study. They have demonstrated that HUP27 with additional woven fabric reinforcements improves in comparison to pure HUP27 with regard to the tensile, flexural, and impact properties (Figure 17) [88].

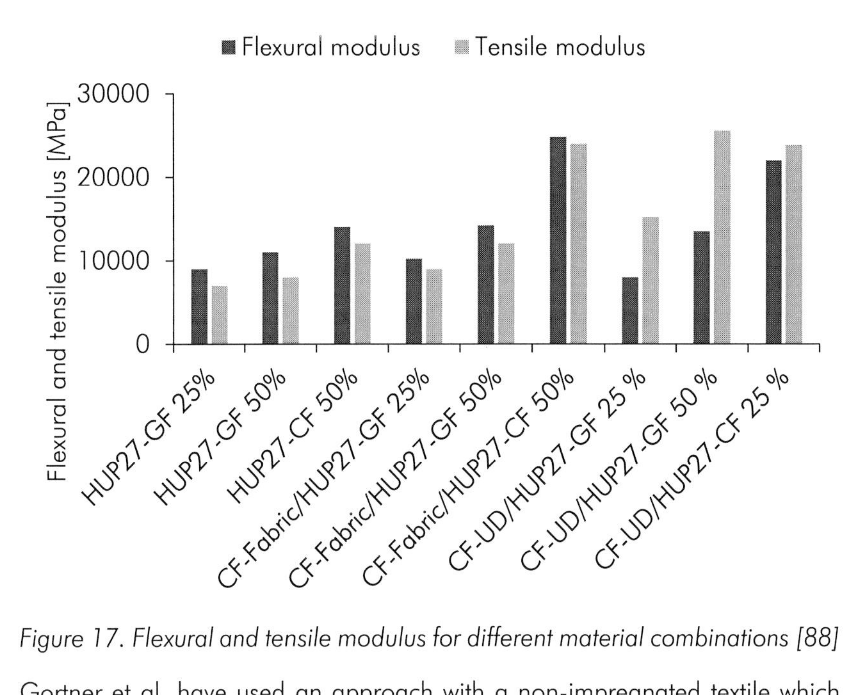

Figure 17. Flexural and tensile modulus for different material combinations [88]

Gortner et al. have used an approach with a non-impregnated textile which was added to the compression moulding of SMC. A non-crimp fabric with glass fibres was used. In their study, they discussed the permeability, the viscosity, the void content, the mechanical bending, and impact properties [89]. The out-of-plane permeability has been determined by a measurement cell. The results have shown a permeability between 10^{-11} and 10^{-13} m² in dependence on the fibre volume content of the non-crimp fabric. They have found out that the viscosity of the SMC material is strongly influenced by the filler content. Higher filler contents result in higher viscosity which led to a poorer impregnation. But higher processing compressions counteracted the increase of the viscosity which is in accordance with Darcy's Law. An increase of the mechanical properties has been evaluated with an increase of the non-crimp fabric layers [89].

2.2.3 Non-fibre reinforcements

Beside typical fibre reinforcements, alternative materials can be combined with SMC materials to manufacture a new kind of hybrid composite.
Rigid foams can be used as core materials to increase the part's stiffness. The surface layers represent the conventional SMC material. A great challenge is the regardful material selection as well as process control because processing

compressions need to be below the allowable compression stresses of the foam. Otherwise, the foam collapses. A significant rise of the flexural properties in comparison to the monolithic materials has been observed. Structural simulations have shown a good agreement with the experimental results regarding the elastic behaviour and the crack initiation [56].
Another option is the hybridization between a SMC material and steel sheets. Joining both materials has been achieved by using an adhesive bonding after compression moulding of the SMC. In contrast to the SMC-foam-technology, the steel sheets represent the surface layers to improve the impact properties. The SMC material is located in the core region. In a recent study, experimental investigations have proven the prediction of higher composite impact resistance in contrast to pure SMC materials. Moreover, a weight reduction of 11 % has been achieved. Furthermore, the specific yield strength was higher than aluminium, stainless steel, as well as high-strength steel [57].
The so-called 'thermoset-overmoulding' process is another way to merge SMC materials with a composite shell. A fully cured fibre reinforced thermoset is laid into an open mould with additional semi-finished SMC material. Afterwards, the compression moulding starts and the high-reactive SMC begins to cure on the surface of the already cured thermoset. Therefore, the thermosetting shell requires an in-depth surface preparation to achieve excellent conditions for the co-bonding between both materials. This technology allows an easy implementation of rips which increase the stiffness of the overall composite part [58].

3 Hybrid Sheet Moulding Compounds

In this study, Hybrid Sheet Moulding Compounds (Hybrid SMC) are characterized by a combination of a SMC material and a reinforcing carbon fibre textile. The carbon fibre textile is made by the Tailored Fibre Placement (TFP) technology. It offers a high degree of flexibility because fibre bundles can change the fibre direction at any time. The reinforcing textile is characterized by no kind of pre-impregnation. The fibre impregnation takes place during the compression moulding, when the resin within the SMC material is transferred to the carbon fibre reinforcement. Figure 18 shows the manufacturing chain to process a Hybrid SMC laminate with a continuous carbon fibre reinforcement.

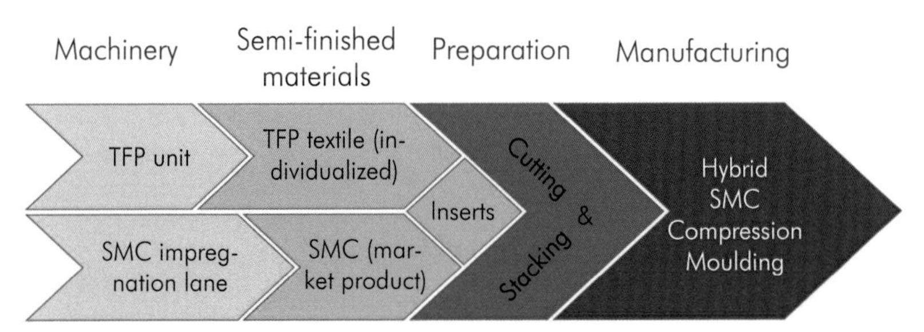

Figure 18. Individual steps to process Hybrid Sheet Moulding Compounds

The SMC has to be cut as well as the TFP textiles. Both semi-finished products have a good buy-to-fly ratio because cutting scraps of the SMC material can be reused in other process cycles. In this context, TFP textiles just use the quantity of high-cost carbon fibre materials which is required for the part performance. The semi-finished SMC and TFP materials are stacked on each other.

It is important that the reinforcing TFP textile is always in the core of the Hybrid SMC because a full impregnation cannot be realized with an exclusive one-sided fibre impregnation. Optionally, metallic inserts for load introduction elements or screw joints can be already implemented. After the stacking of the SMC material and the fibre reinforcement is completed, the material stack is moved to the open and preheated mould inside the hot press. Hereafter, the compression moulding process is started. The implementation of the compression moulding as well as the post-processing do not deviate to the conventional SMC processing. In a previous study, Fette et al. have demonstrated that (semi-)automation of the process chain is implementable for Hybrid SMC composites to fulfil typical production rates for cabin linings of single aisle aircrafts [90].

3.1 Textile Manufacturing

Continuous carbon fibre reinforcements can be added to the compression moulding process to strengthen the SMC composite. Conventional carbon fibre fabrics have high scrap rates and lead to high material costs. Within woven or non-crimp fabrics, the fibre directions are equal at any place of the semi-finished product. In the past, several alternative methods were developed which enable higher flexibilities like Automated Tape Lay-up (ATL), Automated Fibre Placement (AFP), or Tailored Fibre Placement (TFP) [91]. The ATL and AFP technology use pre-impregnated tapes which are laid down on a one-sided mould in any direction of choice. However, the direction of the tape is rigid and cannot be changed anymore after starting the lay-down process. The tape width is the main difference between the ATL and AFP technology. The AFP technology processes plenty of tows up to a width of five inch which allows an improved drapability. In contrast to AFP, the ATL technology uses tapes up to 300 mm [92]. In comparison to the ATL and AFP technology, the TFP technology has a higher degree of design freedom. The production of carbon fibre reinforcements can be made in a previous production step. Furthermore, the manual handling of the carbon fibre reinforcement can be easily performed. Therefore, the TFP technology is used in this study.

3.1.1 Tailored Fibre Placement

Tailored Fibre Placement is a technology based on stitching and allows the implementation of load-path optimized fibre architectures. On the contrary to conventional semi-finished continuous fibre textiles, the TFP technology does not realize rigid fibre orientations within the textile. It is able to change the fibre direction at any time. Therefore, it is a technology with a high degree of design freedom. A stitching unit is required for the manufacturing of semi-finished textiles made by TFP. In this study, the unit JGW 0200-550 by ZSK Stickmaschinen GmbH was used (Figure 19).

Figure 19. TFP unit 'JGW 0200-550' by ZSK used in this study [93]

In a first step, the developed textile design is transferred to a sketch data which is subsequently implemented into the stitching unit. The unit consists of a base table and a stitching arrangement. The stitching arrangement is shown in Figure 20.

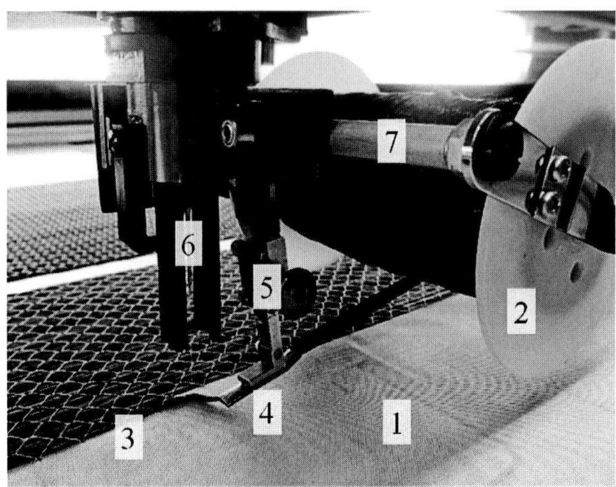

Figure 20. Embroidery head of a TFP stitching unit

On the work space is a clamping device to fix the base material (1). The clamping device is freely movable in the X- and Y-direction and controlled by the stitching unit. Either semi-finished textiles or thermoplastic foils can be implemented as base materials. The embroidery head consists of several components. The continuous fibre bundles are wound on a bobbin (2). The end of the fibre bundle is led through a bundle pipe (4) to fix it with the base material (3). The bundle pipe is fixed on a pendulum (5) which swings continuously from the left to the right and back again. After every change of the pendulum's direction, the needle (6) stitches next to the continuous carbon fibre bundle to fix the fibre bundle on the base material. The bobbin is freely moveable around the needle which allows sudden changes in the fibre direction. The TFP technology allows the complete automatic manufacturing of two-dimensional textiles. The realization of curved shell geometries can be achieved by draping the TFP textile. Modern stitching units work with a production speed above 1000 stitches per minute. High production rates can be achieved in dependence on the complexity of the textile's geometry. Furthermore, the scrap rate of the fibre materials is also low because the high design flexibility allows an ideal use of the fibre material. Since today, the TFP technology is still a niche for the manufacturing of a semi-finished textile. However, there are some series applications in the mechanical engineering and the transportation industry (Figure 21).

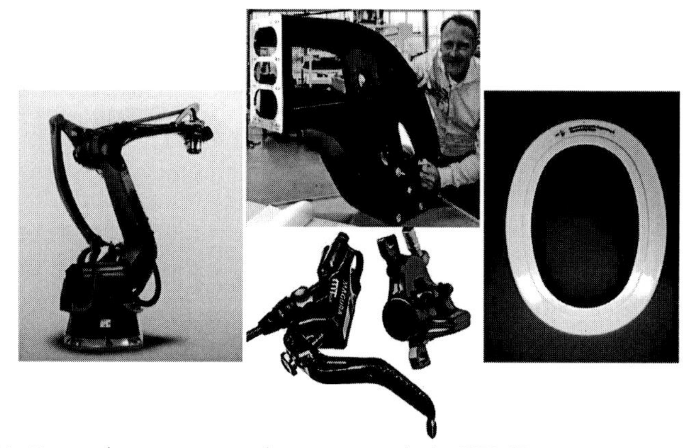

Figure 21. Exemplary series applications made by TFP (Faserinstitut Bremen e.V. was participated at the development of the window belt for the Airbus A350 and the robotic removal system) [94]

Other TFP reinforcements were already developed for bicycle components like a brake booster, a link plate, a bicycle frame [95], or bladed rotors [96].

In this study, a glass fibre woven fabric was chosen as a base material. The product code is 04391 and supplied by Porcher Industries Germany GmbH. It is a plain weave with a surface weight of 105 g/cm^2 which results in a total thickness of 0.095 mm. It is an ideal base material because of the strong shape stability which is not impaired by the needle stitches of the sewing thread. A polyester sewing thread has been used. The product name is Serafil 200/20 and it is supplied by Amann & Söhne GmbH & Co. KG. A high strength carbon fibre has been used for the continuous reinforcing fibre. It is supplied by Teijin Carbon Europe GmbH and is described with HTS45 F13 with a yarn count of 800 tex.

3.1.2 Classification of TFP Textiles

To systematically investigate the fibre impregnation of a TFP textile by a SMC material, a classification for the highly flexible fibre design is required. A frequent application for TFP reinforcements are load introduction areas. By using conventional semi-finished textiles as reinforcements, the load-carrying fibres are damaged and lead to a significant reduction of the mechanical properties. By using the TFP technology for the design of a load introduction area, loop structures are an appropriate fibre design for reinforcements. Mechanical loads are perfectly absorbed by the form closure and transferred along the unidirectional fibre bundles into the structure. Crothers et al. have implemented TFP reinforcements in open-hole tension plates. According to the TFP design, the specific tensile strength was increased by 45 % and 68 %, respectively [97]. Schürmann has described the loop joint as a joining technology with the best mechanical performance in combination with the lowest weight when punctual load occur [22]. According to a loop structure, this study uses two main fibre morphologies. Usually, load introducing areas use a curved fibre architecture and a load transferring areas use a elongated fibre architecture (Table 2).

Table 2. Classification of loop structures made by TFP

Fibre morphology	Function
Elongated fibre architectures	Load transfer
Curved fibre architectures	Load introduction

As an example for processing TFP textiles with the specific fibre design, Figure 22 shows an aircraft bracket by using a hybrid method of construction because it combines CFRP as well as titanium elements [98]. The carbon fibre reinforcement was manufactured by using TFP and afterwards impregnated by an epoxy resin. It illustrates the use of elongated and curved fibre architectures. Curved fibre architectures are used in the areas of load introduction. In this case, mechanical loads are introduced at the pin joint and transferred with an elongated fibre architecture to the metallic flange. The flange is characterized by a curved fibre design at the angles.

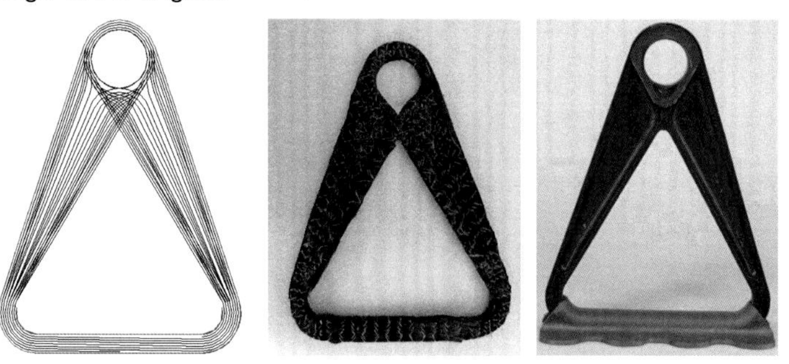

Figure 22. Bracket element: stitching paths, TFP textile, final (hybrid) composite part

Elongated fibre architectures

The implementation of straight fibre architectures can be realized by the TFP technology. However, small deviations within the fibre straightness lead to undulations and therewith a reduction of mechanical properties [99]. Due to the pendulum movements, the continuous fibre bundle slightly moves to the side at the same time which finally leads to in-plane undulations. In addition, fibre misalignments can occur with an increasing stitch length d_{sl} (Figure 23, left) [100]. But also out-of-plane undulations occur while the fibre lay down as well due to the sewing thread, which locally narrows the continuous fibre bundle down to the base material. Due to this constriction of the carbon fibre bundle, the local fibre volume content locally increases (Figure 23, right) [101].

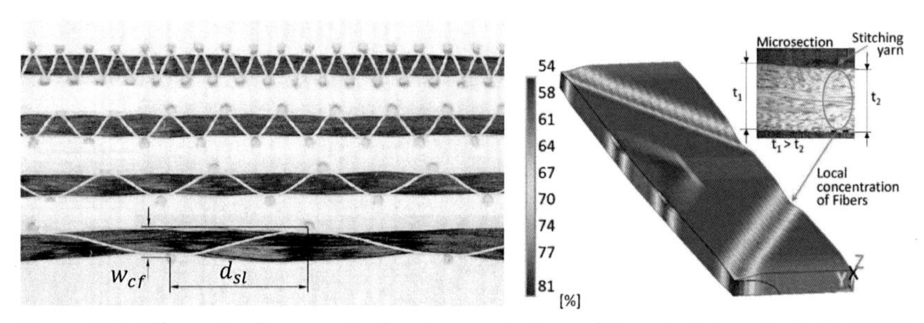

Figure 23. Fibre misalignments due to a variation of the stitch length (left) [100] and constriction of the carbon fibre bundle due to the sewing thread (right) [101]

Both effects can be reduced by adapting the TFP parameters. Additional fibre prestress during the composite manufacturing leads to a reduction of the in-plane undulations. Less stress within the upper thread would lead to a decrease of the fibre bundle constriction and therefore a decrease of the out-of-plane undulations but little fibre bundle constrictions always still remain.
In this study, elongated fibre architectures are implemented by unidirectionally arranged continuous carbon fibre bundles (Figure 24).

Figure 24. Elongated continuous carbon fibre bundles made by TFP

After each unidirectional fibre arrangement, a turnaround of 180 degree follows and the continuous fibre bundle lays next to the neighboured fibre bundle with the bundle distance d_{FB}. The length as well as the width of the unidirectional TFP textile is equal to the mould. The complete mould is covered with the TFP reinforcement. Therefore, sideward deformations of single carbon fibre bundles, which can occur during compression moulding, can be neglected.
To increase the amount of carbon fibre reinforcements, further layers of unidirectionally arranged carbon fibres can be directly laid down on the layer beneath. There are three methods to lay down continuous carbon fibre bundles by a TFP unit, especially when carbon fibre layers are stitched on each other (Figure 25). The first design is characterized by an identical position of each carbon fibre bundle from the second on the first layer (Figure 25, left). In contrast to the first design, the carbon fibre bundles with the second design exactly lay between the carbon fibre bundles of the previous fibre layer. (Figure 25, middle). Then, the fibre bundle distance is equal to the width of the fibre bundle. The third design is characterized by shorter bundle distances than the width of on carbon fibre bundle. Thus, the carbon fibre bundles partially lay on the neighboured fibre bundle (Figure 25, right). Concerning the homogeneity of

thickness, the second and the third TFP design have shown better results, than the first TFP design. The main reason is heterogeneous thickness of a single carbon fibre bundle due to its oval shape. In addition, the presented effect of undulations intensifies, when the sewing thread is always on the same line [102].

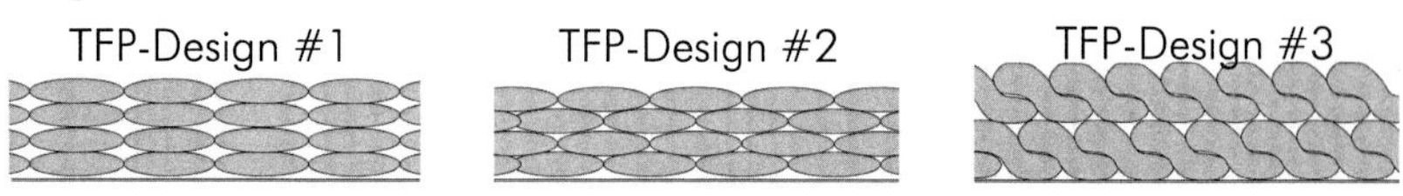

Figure 25. Designs for processing unidirectionally arranged carbon fibre bundles

Investigations can be performed with a qualitative and quantitative evaluation of the result. To develop an analytical model to predict the fibre impregnation, a quantitative method is required. Therefore, the elongated fibre architectures are investigated in a *quantified* way.

Curved fibre architectures

Usually, curved fibre architectures occur in TFP textiles when external loads are introduced into the composite structure. In these cases, the fibre design of the reinforcements is essential for the overall mechanical performance. Schürmann has already explained that punctual and high tensile loads can be excellently absorbed with an unidirectionally arranged continuous carbon fibre reinforcement design which entangles the point of load introduction [22]. The TFP technology allows a simple implementation of curved fibre architectures. However, the fibre bundle tends to lay at the inner area of the fixing sewing thread. This effect results in gaps between the fibre bundles (Figure 26). A reduction of the fibre bundle distance d_{fb} in curved areas would fix the problem. The stitch density increases at curved fibre architectures to ensure the position of the fibre bundle.

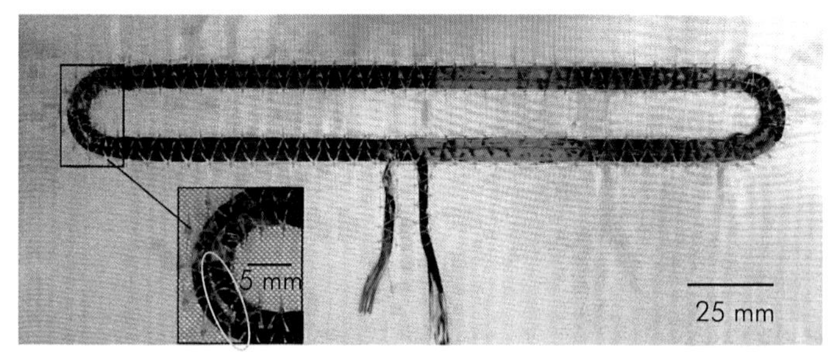

Figure 26. Gaps between the fibre bundles at curved fibre architectures

Metallic inserts are used to offer the continuous carbon fibre reinforcement a lateral support to avoid any sideward deformation of the fibre reinforcement which leads to an increase of the mechanical performance [22]. Considering the fibre impregnation of curved fibre geometries, the main difference is the direction of the reinforcement's impregnation. Metallic inserts hinder the fibre impregnation through the thickness (in contrast to the elongated fibre architectures) that is why the impregnation is implemented in-plane (Figure 27). Especially the impregnation of the inner fibre bundles is of great importance because the inner fibre bundles absorb most of the occurring loads.

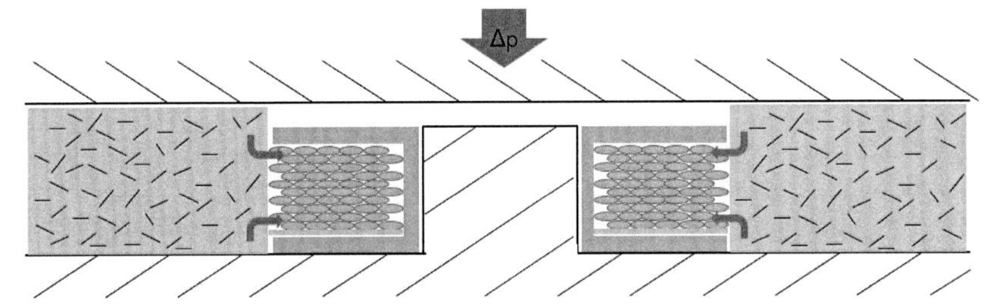

Figure 27. Resin flow during the compression moulding of Hybrid SMC with metallic inserts

The in-plane fibre impregnation is dependent on several aspects like the friction between the SMC material and the mould or the metallic insert, the processing compression, the thickness of the carbon fibre loop, the fibre prestress, and the stitch density for the curved fibre architecture. For the investigation of curved fibre architectures, the same aluminium mould is used. Beside the manufacturing of flat laminates, the mould can be equipped with cylindrical tooling inserts (Figure 28). The metallic elements as well as the textile loop structures can be fixed with these cylindrical tooling inserts.

Figure 28. Flexible aluminium mould for the manufacturing of Hybrid SMC

The investigation of an impregnation model for elongated fibre architectures already requires a comprehensive investigation. Due to the high number of new and important parameters (fibre prestress, insert diameter) and conditions (in-plane impregnation), the investigations on the curved fibre architectures does not meet the circumstances of the elongated fibre architectures that is why a total transfer of the impregnation model cannot be implemented. Therefore, experimental investigations restrict on an illustration of the complexity of the topic. The evaluation takes place in a *qualitative* way.

3.2 Compression Moulding

Compression moulding of SMC materials with a non-impregnated continuous carbon fibre textile is almost similar to the conventional compression moulding process, however it contains some differences. The continuous carbon fibre reinforcements are always placed in the core region of the Hybrid SMC, otherwise a complete fibre impregnation cannot be realised. Taking this requirement into account, both one (Figure 29) or two plies of the TFP reinforcements are used in the experimental tests and placed in the middle of the compression moulding process. By using two TFP reinforcements, two base materials are included within the Hybrid SMC composite. Then, it is recommended to put the base material on the transition area of the continuous carbon fibre bundles and the SMC cover layer.

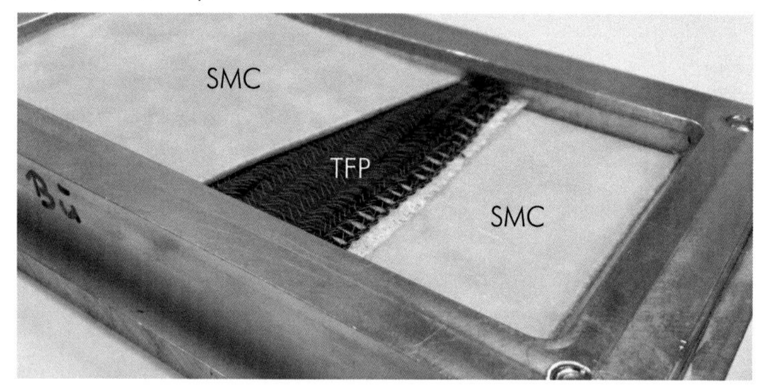

Figure 29. Stacking sequence of Hybrid SMC

The mould is characterized by a rectangular shape with a radius of 5 mm at the corners. The length of the mould is 250 mm and the width is 120 mm. The blank size of the SMC material is chosen to 95 % of the overall mould area. It is higher in comparison to conventional compression moulding processes of SMC (60 – 80 % [43]). On the one hand, shear stresses on the continuous carbon fibre reinforcement with its effects should be avoided. On the other

hand, a blank size below 100 % allows the compression moulding to squeeze the enclosed air out of the mould. In addition, the fibre impregnation reduces to a one-dimensional through-thickness resin flow. This constraint will be considered in the analytical impregnation model.
In this study, the SMC material is supplied by Polynt Composites GmbH and has the identification code HUP63/25-RN1090. The SMC is based on an unsaturated polyester resin which cures due to heat supply. The reinforced fibres are glass fibres in the shape of long fibres with a fibre length of 25 mm. The weight content of the glass fibres is 25 % which is equal to a fibre volume content of 22 %. Filler materials are one of the major constituents. Especially, the SMC material HUP63 has a high content of aluminium hydroxide (ATH). It ensures excellent flame retardant properties wherefore HUP63 passed the required certification test for the use in aircraft's cabin area [103]. The material data sheet of HUP63 recommends a processing temperature between 135 °C and 155 °C and a processing compression between 80 bar and 120 bar. The curing time depends on the part thickness which is 25 seconds per millimetre. Certainly, the process recommendations refer to the compression moulding of pure SMC material. The individual process parameters can deviate from the recommendations of the supplier with additional continuous carbon fibre reinforcements.

3.3 Characteristics of a Hybrid SMC Composite

This chapter deals with the characteristics of the finished Hybrid SMC composite with the integrated continuous carbon fibre reinforcement. The presented characteristics can be divided into the microscopic and the mechanical properties. The Hybrid SMC composites are mainly investigated with a microscopic analysis of the polished cross-sections. Here, results concerning the fibre volume content as well as the void content can be achieved. In addition, an analytical method is presented which calculates the fibre volume content within the SMC surface layers after compression moulding because it changes due to the resin transfer into the carbon fibre reinforcement. Afterwards, the mechanical properties of Hybrid SMC composites with continuous carbon fibre reinforcements are described. Trauth has exclusively worked on the mechanical properties of Hybrid SMC composites with non-crimp fabrics as carbon fibre reinforcements. Her results are summarized to illustrate the most important mechanical parameters [104].

3.3.1 Microscopic classifications

To give an overview of the constituents in Hybrid SMC composites, a description is made based on microscopic images. Therefore, a piece of a fully cured

Hybrid SMC composite was cut out perpendicular to the fibre direction andprepared to a polished cross section for the microscopic analysis. An image section is shown in Figure 30.

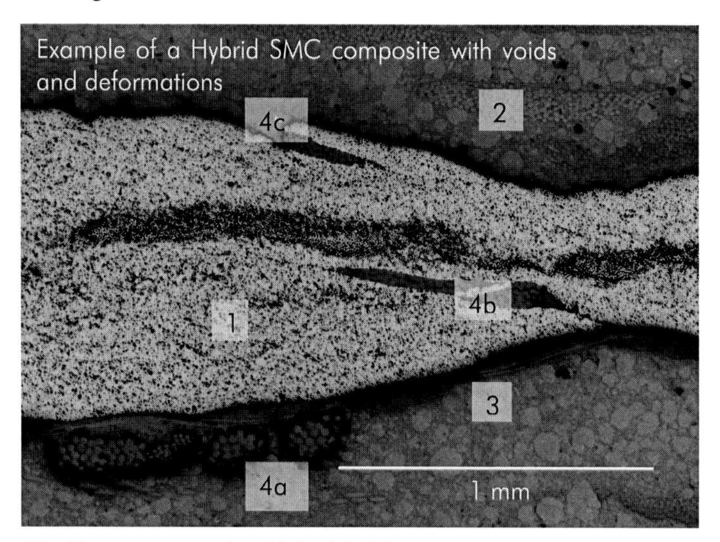

Figure 30. Constituents within Hybrid SMC composites

The constituents of the SMC material as well as the TFP textile can be clearly identified. The continuous fibre reinforcement **(1)** is characterized by the small carbon fibre diameter. It is located in the middle of the Hybrid SMC. The continuous carbon fibre textile is surrounded by the SMC material **(2)**. Between the carbon fibre bundles and the SMC material, a black shadow can be noticed. This is a transition area between the hard carbon fibre bundles and the soft SMC material. Therefore, it is not a defect at the interface between the carbon fibre reinforcement and the SMC material but a normal response due to material polishing of the difference hardness levels of the material. Furthermore, the single components of the SMC materials are clearly visible. Most of the glass fibres are clustered in fibre bundles. The bundles are impregnated by the UP resin. In addition, some conglomerates with non-uniform dimensions were identified inside the SMC material which are filler constituents. One side of the carbon fibre reinforcement is characterized by the glass fibre woven fabric which represents the base material **(3)**. The undulating fibre orientation clearly catches the eye. The carbon fibre bundles are fixed on the base material with a sewing thread. Within a polished cross section, the sewing yarn can be located below the base material **(4a)**, inside the carbon fibre bundles **(4b)**, and on top of the carbon fibre bundles **(4c)**.

However, at first glance, Figure 30 does not show a homogeneous structure within the cured Hybrid SMC. On the one hand, voids are formed in the middle of the continuous fibre reinforcement and lead to no sufficient fibre impregnation. In contrast to the middle, the edge areas represent better impregnation results. Curing reactions have taken place before the total fibre impregnation has been completed. On the other hand, two different fibre volume contents were adjusted within Hybrid SMC composite. First, a high fibre volume content is noticed within the carbon fibre bundles. The second fibre volume content results from the SMC material with its containing glass fibres. But the fibre volume content changes during compression moulding due to the resin transfer from the SMC material into the continuous carbon fibre reinforcement. A reduction of resin leads to an increase of the fibre volume content within the SMC cover layers. The analysis of the fibre volume content of the continuous carbon fibre reinforcement as well as the glass fibres within the SMC material is investigated in the following chapter. Moreover, local changes of the reinforcement's thickness can be noticed. This is an effect for the use of an inappropriate design during the manufacturing by TFP (Figure 25) [102].
Figure 31 shows images which were taken with the scanning electron microscope (SEM) EVO 10 by ZEISS. Both images show the transition area between the SMC cover layers and the carbon fibre reinforcement including the glass fibre base material and the polyesters sewing thread.

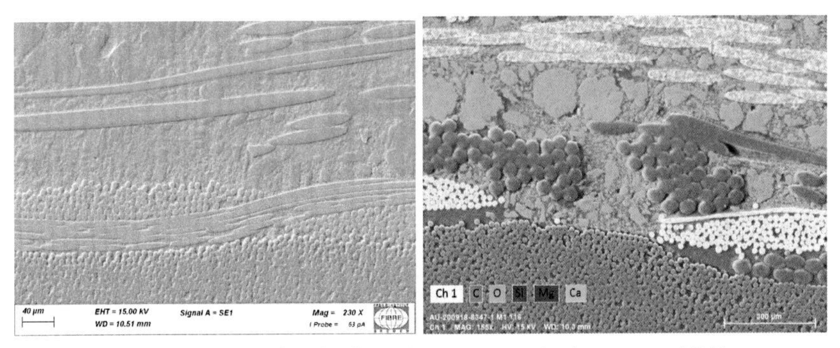

Figure 31. SEM images of Hybrid SMC composite (right image: EDX)

The images were made with a 230 times magnification and show the heterogeneous composition of the SMC material. The left image proves the impregnation of each carbon fibre filament at the transition area by the SMC material. In contrast to the left image, the right image shows the individual chemical components of the Hybrid SMC composite by using the energy dispersive x-ray (EDX) microanalysis. By using this method, the energy of the occurring x-rays is measured. Each chemical element sends a characteristic x-ray signal which is triggered by the primary electron beam and finally results in an element mapping. A high percentage of calcium was found within the SMC cover layers which is a standard filler material. Furthermore, the existence of magnesium was proven which is important for the increasing flame-retardant properties in the form of magnesium oxide.

3.3.2 Decomposition effect

The fibre volume content is a crucial characteristic regarding the mechanical properties of FRP. In Hybrid SMC composites, two various fibre volume content occur. High processing compressions emboss the fibre volume content of the continuous carbon fibre reinforcement during the fibre impregnation. The determination of the fibre volume content within the continuous carbon fibre reinforcement will be discussed in Chapter 5.1.3. The fibre impregnation is achieved by the resin within the SMC material which is partially transferred due to the compression moulding. Here, a decomposition takes place. The reduction of the resin within the SMC material leads to an increase of the fibre volume content within the SMC cover layers. The glass fibres as well as the filler materials are not affected by the resin transfer because the polished cross sections (e.g. Figure 30) do not show any evidence. Therefore, the glass fibres still remain in the SMC surface layers.
The increase of the fibre volume content within the SMC material can be explained with an analytical approach. The analytical approach is based on the microscopic images of the specimens' cross-sections which is also used within the experimental design (Chapter 5.1.2). The initial fibre volume content of the SMC is 22 %. It was calculated by the glass fibre weight content with the following equation:

$$\varphi_{GF} = \frac{1}{1 + \frac{1-\psi_{\mathrm{F}}}{\psi_{\mathrm{F}}} \frac{\rho_F}{\rho_{SMC}}} \quad . \tag{3.1}$$

The determination of the fibre volume content in the SMC surface layers can be described by the following inversely proportional dependency regarding the cross section of the Hybrid SMC composite:

$$\varphi_{GF,new} = \frac{A_{SMC}}{A_{SMC,new}} \varphi_{GF} \quad . \tag{3.2}$$

A_{SMC} is the area of the SMC before the compression moulding (cross-section of the semi-finished SMC) and $A_{SMC,new}$ is the area of the SMC cover layer after compression moulding. The areas can be reduced to the thickness of the investigated sections which leads to following equation:

$$\varphi_{GF,new} = \frac{z_{SMC}}{z_{SMC,new}} \varphi_{GF} \quad . \tag{3.3}$$

The thickness of the semi-finished SMC z_{SMC} can be calculated with the surface weight Q_{SMC} and the density ρ_{SMC} from the material data sheet:

$$z_{SMC} = \frac{Q_{SMC}}{\rho_{SMC}} \quad . \tag{3.4}$$

After the compression moulding was completed, the thickness of the SMC cover layers can be measured by using the microscopic images of the polished cross sections. It is recommended to use the average thickness within the cross section to consider slight deviations within the thickness. The final results show a slight increase of the fibre volume content in the SMC surfaces layers. The fibre impregnation of a TFP textile with four layers of continuous carbon fibre reinforcement, which was equal to a reinforcement's thickness of 0.7 mm, leads to a fibre volume content of 23.53 %. This is an increase by 1.53 % to the initial 22 %. The experimental study has been conducted with a SMC material which has a surface weight of 4.000 g/m^2. Lower surface weights could either lead to a stronger increase of the fibre volume content in the SMC surface layers or to an insufficient impregnation of the continuous carbon fibre reinforcements. All in all, this section has presented a calculation for the determination of the fibre volume content within the SMC surface layers after compression moulding. Image analysis in combination with an analytical approach has been used. A decomposition effect was noticed within the SMC surface layers because the SMC material represents a resin carrier which is transferred to the continuous carbon fibre reinforcement. Therefore, the glass fibre volume content slightly increases but it is influenced by the SMC content within the overall composite.

3.3.3 Mechanical Properties

This chapter presents the mechanical properties of the SMC composites made by HUP63 as well as the mechanical properties of Hybrid SMC composites. For this purpose, Hybrid SMC composites of the IRTG were used for this illustration because they have considered a consistent continuous fibre reinforcement. The results cannot be directly transferred to the Hybrid SMC composites in this study because of two reasons. First, in contrast to consistent non-crimp fabrics local fibre reinforcements, which were used by the IRTG, this study considers the TFP technology which is used to design local reinforcements. Second, IRTG uses a different type of SMC material with another composition of fillers which has an impact on the mechanical properties. Nevertheless, the improvement of the mechanical properties due to carbon fibre reinforcements can be illustrated.

The SMC material: HUP63/25-RN1090

This chapter describes the material properties of the pure SMC material HUP63. The material data sheet of the pure HUP63 gives information about the flexural modulus and the flexural strength. The flexural modulus of the material is described with 9,000 MPa and the flexural strength is described with 135 MPa. Experimental tests according to DIN EN ISO 14125 have confirmed the flexural properties of the declared process recommendations. The flexural properties in production direction (0° PD) are slightly higher (Figure 32).

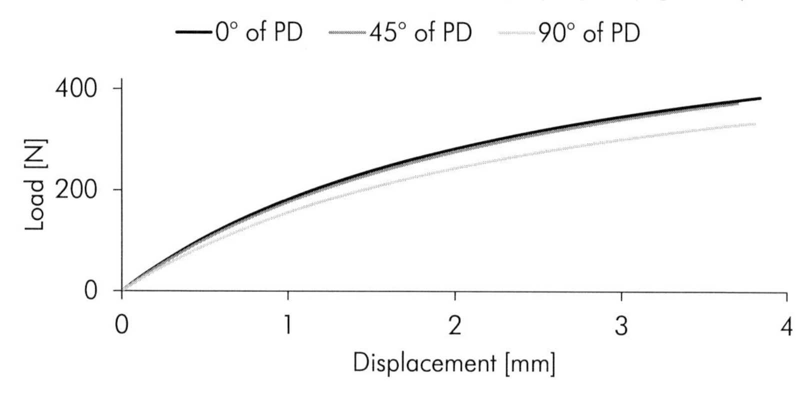

Figure 32. Flexural properties: Load-displacement-diagram for HUP63

The tensile strength has been determined in tensile tests according to DIN EN ISO 527. The tensile strength is 60.7 MPa for tested specimens in production direction of the semi-finished SMC material. The tensile tests have shown a clear non-linear failure behaviour like the flexural tests have shown as well. This failure behaviour can be described by an empirical equation found by Kästner [105], with the tensile modulus E, the strain ε, and the empirical parameter β:

$$\sigma = \frac{E}{1+\beta|\varepsilon|}\varepsilon \quad . \tag{3.5}$$

This has already been demonstrated in the past for the SMC material HUP63 [106]. The stress-strain-diagram can be seen in Figure 33. The maximum stress is relative independent to the production direction in contrast to the maximum load of the flexural properties (Figure 32). Only the strain becomes higher.

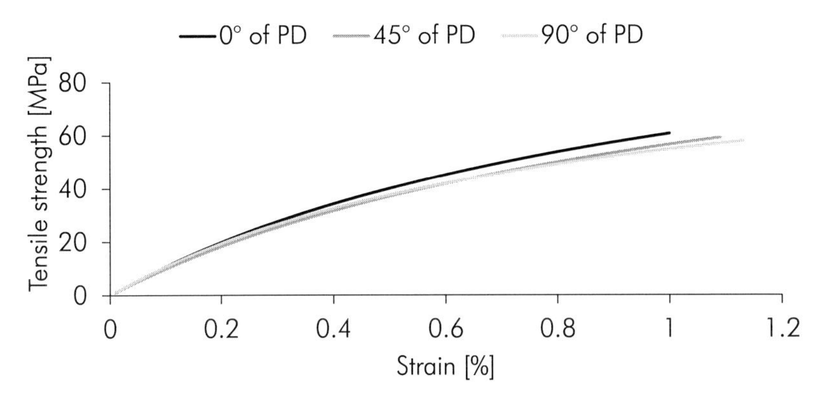

Figure 33. Tensile properties: Stress-strain-diagram for HUP63

The experimental results regarding the flexural and tensile properties with the standard deviations (SD) are summarized in Table 3.

Table 3. Flexural and tensile properties of HUP63

	Flexural properties			Tensile properties		
	Strength [MPa]	Modulus [MPa]	SD [%]	Strength [MPa]	Modulus [MPa]	SD [%]
0°	138.6	11,516.9	7.8	62.6	12,058.7	3.5
45°	134.5	11,481.2	8.7	61.3	11,520.7	6.3
90°	128.1	10,818.6	10.1	60.7	11,617.9	5.1

Overall, the mechanical properties are limited in comparison to other SMC materials. Two reasons were identified. First, a relative low glass fibre volume content was used. Second, the limitation can be a result of the high filler content. A standard deviation of up to 10 % have been found for other SMC materials in the literature as well. This can be explained by the significant fabrication-induced heterogeneity of the material [107].

Hybrid SMC containing continuous fibre reinforcements

In this study, the TFP textiles are classified into elongated and curved fibre architectures. TFP textiles with elongated fibre architectures have a high resemblance with unidirectional non-crimp fabrics. Both the unidirectionally arranged continuous carbon fibre bundle and the fixing sewing thread are essential elements of the reinforcing structure. This section refers to Trauth who has used impregnated non-crimp fabrics in combination with SMC materials (CoDi-

CoSMC) [104]. In her investigations, she has comprehensively studied the mechanical properties of CoDiCoSMC [104] but there is an important difference to the Hybrid SMC composites in this study. The CoDiCoSMC composites were characterized by a different stacking sequence. The core layer was a long fibre reinforced SMC material and the surface layers were an unidirectional and pre-impregnated continuous carbon fibre reinforcement (CCFR).

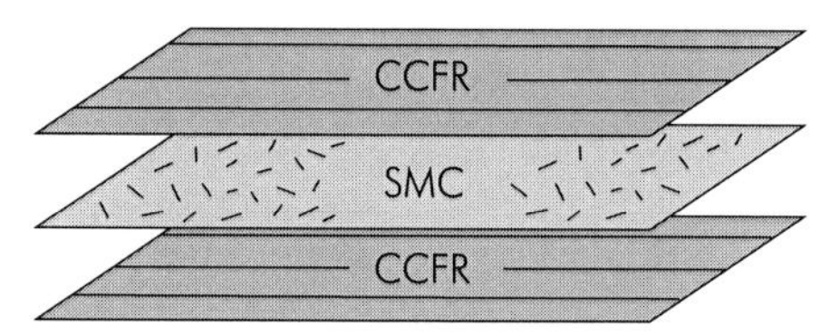

Figure 34. Stacking sequence of a CoDiCoSMC

This stacking sequence has been chosen because of the already existing fibre impregnation of the reinforcements. Therefore, it does not require a fibre impregnation anymore. Furthermore, this stacking sequence improves the flexible properties in comparison to the stacking sequence in this study. The comparison with the pure SMC material allows an insight of the impact of the carbon fibre reinforcement. Beside the tensile and flexural properties, the compression and impact properties of Hybrid SMC have been determined as well. The results are summarized in Table 4.

A strong increase of the tensile, flexural, and compression properties can be seen. The highest growth can be noticed for the tensile strength and the flexural modulus due to the continuous carbon fibre reinforcement. Only the material's elongation decreases during the tensile tests. In addition, it has been shown that a modified rule of mixture is able to predict the resulting stiffness in CoDiCoSMC [9].

Table 4. Comparison of mechanical properties of CoDiCoSMC (0°) with the SMC material without any reinforcements [104]

	Unit	SMC	CoDiCoSMC	Percentage
Tensile modulus	[GPa]	11.8	36.3	+ 208 %
Tensile strength	[MPa]	158.0	532.0	+ 237 %
Elongation	[%]	1.6	1.5	- 0.1 %
Flexural modulus	[GPa]	12.6	59.3	+ 371 %
Flexural strength	[MPa]	283.0	570.0	+ 101 %
Compression modulus	[GPa]	12.6	31.7	+152 %
Compression strength	[MPa]	298.0	302.0	+ 1 %

4 Approach

In the past, the Darcy equation has become well-established for FRP as an analytical calculation. It describes the flow of a fluid through a porous medium [108]. In detail, the Darcy equation links the permeability K, the material's viscosity η and the pressure gradient $\frac{\Delta P}{\Delta z}$ with a linear dependency which finally results in the flow rate V of the fluid:

$$V = \frac{K}{\eta}\frac{\Delta P}{\Delta z} \quad . \tag{4.1}$$

The Darcy equation is coupled on several boundary conditions [30]. Taking the Hybrid SMC compression moulding process into account, an examination of the individual boundary conditions is done. First, a creeping flow have to exist to use the Darcy equation. Creeping flow exists, when the Reynolds number is below one. The Reynolds number is dependent on the viscosity, the density, and the flow velocity. Materials with a high viscosity, which is assumed for the SMC material HUP63, tend to a creeping flow. Second, the fluid is incompressible which describes that the fluid does not change its density. Slight deviations of the density occur due to the resin transfer into the carbon fibre reinforcement by processing Hybrid SMC composites but the deviations do not have a big impact. The glass fibres remain in the SMC cover layers and lead to the resulting fibre-resin-ratio. Third, the permeability does not change during fibre impregnation. It is supposed that the start of the compression moulding leads to two subsequent actions. After the compression moulding starts, the fibre reinforcement is completely compressed to the final fibre arrangement. Then, the fibre impregnation takes place and the fibre reinforcement does not change its structure anymore [109]. Therefore, a constant permeability is supposed. Fourth, there is no capillary action within the fibre structure. Only the processing compression is the driving force. Without any processing compression, the fibre impregnation would not be realized. In the past, the application of Darcy's Law has already been implemented for the injection compression moulding process [110] and the compression RTM process [111] which are both driven beside the injection pressure by a hot press. Fifth, an isothermal process is required. The compression moulding of SMC fulfils this requirement but the viscosity changes due to crosslinking. The sixth boundary condition is the existence of a Newtonian behaviour of the fluid. In general, SMC materials do not behave like a Newtonian fluid. However, a modification of the Darcy equation can be implemented to consider the non-Newtonian behaviour as well as the fast curing mechanisms of SMC materials within Hybrid SMC compression moulding. Therefore, the modification has to be done on the material's viscosity with regard to the temperature and the time. Taking this modification into account, the modified Darcy equation for Hybrid SMC composites is the following:

$$V(t,T) = \frac{K}{\eta(t,T)} \frac{\Delta P}{\Delta z} \quad . \tag{4.2}$$

Figure 35 illustrates the approach for the development of an impregnation model for Hybrid SMC composites. The development is divided into three parts: The SMC analysis, the textile analysis, and the process analysis. The analysis of the SMC refers to the time- and temperature-dependent viscosity $\eta(t,T)$. The textile and process analysis describe the permeability K, and the pressure gradient $\frac{\Delta P}{\Delta z}$, respectively.

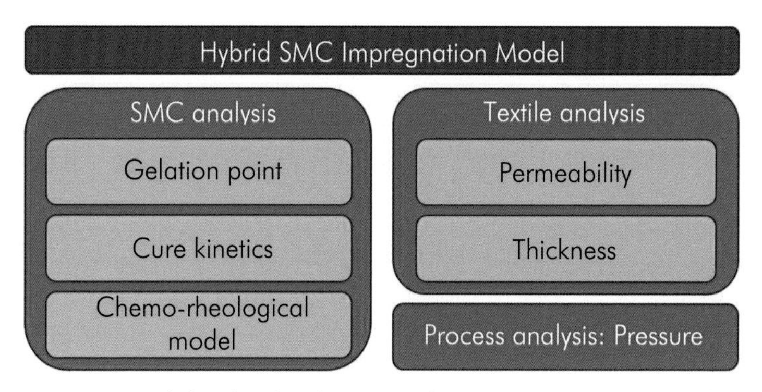

Figure 35. Approach for the development of an impregnation model for Hybrid SMC composites

4.1 Chemo-rheological Model

Chemo-rheology is the viscoelastic behaviour of reacting systems [112]. It is used for the description of the rheological behaviour of cross-linking polymers [113], [114].
An important point during the curing of thermosets is the gelation point. The gelation point means the construction of the first eternal molecular chain within the polymer [115]. Fibre impregnation can only take place before the gelation point is passed [81]. Therefore, the determination of the gelation point is of great importance and therefore the first step. In the next step, a cure kinetic model is developed. The cure kinetic model describes the time- and temperature-dependent cure behaviour. Taking the Hybrid SMC compression moulding into account, the cure kinetic between the start of the curing cycle and the gelation point has to be found but the total enthalpy has to be in common with the results of the material characterization. After the development of the cure kinetic model, the chemo-rheological model can be developed. Here, the rheological behaviour has to be determined taking the cure kinetics into account. Again, only the viscosity between the start of the curing cycle and the pass of gelation point is investigated. The development of the chemo-rheological model is completed by the determination of time-and temperature-dependent viscosity function.

4.1.1 Gelation Temperature

From a chemical perspective, the point of gelation is defined as the first endless molecular network which is built by cross-linking reactions of the thermoset. It means that no more fibre impregnation can take place anymore [115]. Therefore, the gelation point is an important transition area during the fibre impregnation by a thermoset material. The gelation point of individual SMC materials has already been determined by several researchers. Fan and Lee have shown that the degree of cure at the gelation point is less than 1 % [116]. Huang et al. have found that the gelation point is below 5 % of the degree of cure and independent by the processing pressure [117]. This is in accordance to Yousefi and Lafleur who have demonstrated that higher pressures have no impact on the gel conversion but higher processing pressures prolonger the gelation time [118]. De la Caba et al. have found gelation points below 5 % of the degree of cure [119]. Kenny and Opalicki have investigated a degree of cure of 2.4 % at the gelation point in their investigated unsaturated polyester resin [120]. The filler content is an essential parameter for the determination of the gelation point. An increase of fillers decreases the point of gelation [121]. In a study of Halley, he has found out for a SMC material that the gelation conversion is 2.2 % [122]. Overall, past studies have shown that the gelation point is at an

early stage of the curing progress. In contrast to SMC materials, epoxy resins have higher gelation points [123].
The gelation point can be found by the determination of the intersection from the storage modulus and the loss modulus. The ratio of the storage modulus and the loss modulus is called loss angle. It is equal one at the point of intersection [124]:

$$\tan\delta = \frac{G''}{G'} = 1 \quad . \tag{4.3}$$

The determination is based on viscosity measurements by a rheometer. The measurements were done with the oscillated rheometer unit AR2000 by TA instruments. The measurements can be isothermally or non-isothermally controlled. Regarding the determination of the gelation point, non-isothermal measurements are preferred because of the fast curing mechanisms of SMC materials. Here, heating rates of 2.5, 4.0, 5.0, 6.0, 8.0 and 10.0 K/min were chosen up to a maximum temperature of 200 °C. The frequency was set to 1 Hz and the deformation was set to 1.0 %. Figure 36 shows the development of the storage and loss modulus of the SMC material HUP63 with a heating rate of 2.5 K/min but other heating rates have shown a similar behaviour.

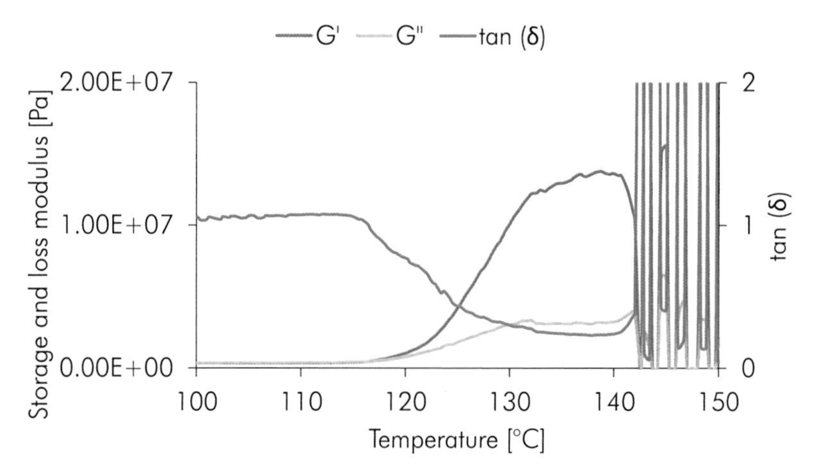

Figure 36. Storage and loss modulus as well as tan (δ) of HUP63 with heating rate of 2.5 K/min

In the beginning of the measurement, the storage as well as the loss modulus are on the same level. A viscoelastic material behaviour is existent. Above a temperature of 115.0 °C, the storage and the loss modulus increase but the storage modulus has a greater gradient. Here, the material cures and becomes solid. Above a temperature of 142 °C, the storage and loss modulus as well as $\tan\delta = 1$ do not follow a clear trend. This unclear response is caused by the oscillating movements of the rheometer and the fast curing properties of the material at the same time. At first glance, a point of intersection is not noticeable because storage and loss modulus are close to each other. However, a detailed view indicates the existence of an intersection ($T_{gel} = 116.4\ °C$) (Figure 37).

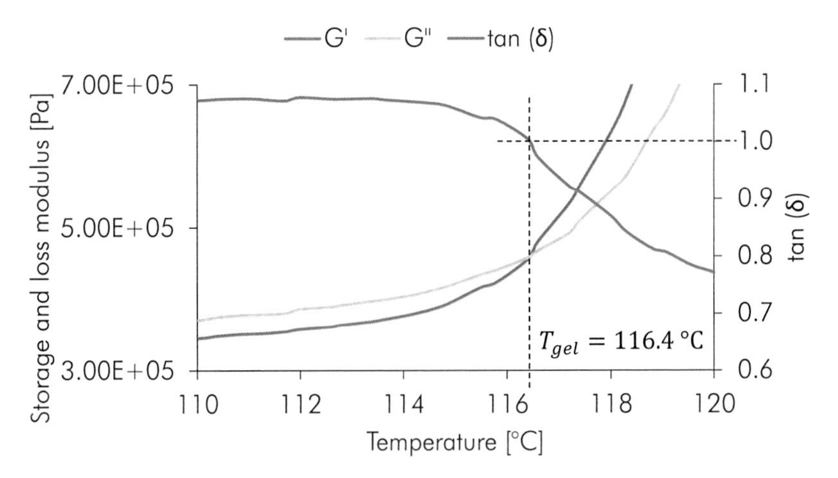

Figure 37. Intersection of storage and loss modulus

The other heating rates were tested analogously. At least, three different measurements were done with each heating rate. The gelation point and the gelation temperature are summarized in Table 5.

Table 5. Determination of the gelation temperature for different heating rates

Heating rate [K/min]	T_{gel} [°C]	t_{gel} [s]	Standard deviation [%]
2.5	116.40	2273	1.50
4.0	119.45	1447	2.90
5.0	119.75	1161	1.59
6.0	120.60	898	0.00
8.0	124.45	761	0.04
10.0	125.70	616	3.19

The experiments have shown that the gelation temperature increases and the gelation time decreases with an increasing heating rate. In the next step, the gelation temperatures are transferred to the cure kinetic model for the SMC material to determine the gelation point in dependence on the degree of cure. The development of the cure kinetic model is described in the following section.

4.1.2 Cure Kinetics

The transformation of a thermoset from a liquid to a solid state is called curing the resin. The progress can be described by the cure kinetics. There are two available methods to investigate the curing progress of a SMC material. These methods are the dielectrically analysis (DEA) and the differential scanning calorimetry (DSC). In comparison to the DSC method, the DEA method has two advantages. First, the measurements are done with real time manufacturing conditions. Second, a high data recording rate allows the recording of the first cross-linking reactions [125]. The method uses a dielectric sensor which is placed in a mould to measure the movement of the charged particles during the curing progress. The charged particles are characterized by the ionic viscosity η_{ion} [126]. In an experimental test, the change of ionic viscosity was investigated with the SMC material HUP63 by using a processing compression of 80 as well as 120 bar and a temperature of 135 °C (Figure 38). The analysis was done with a frequency of 1000 Hz which has shown the clearest signals.

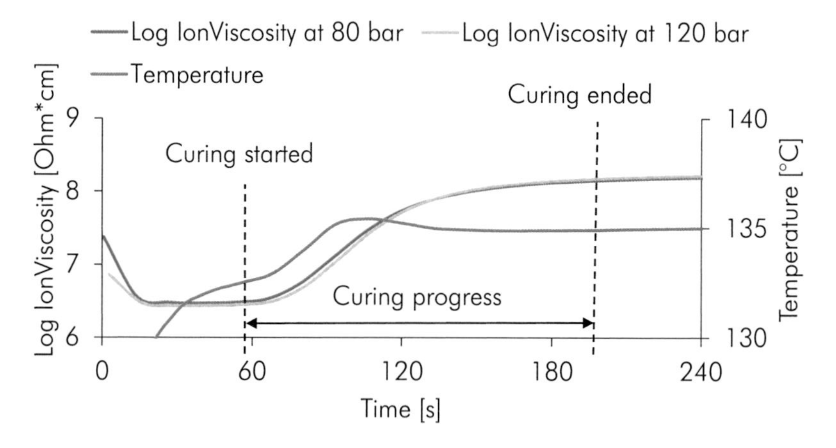

Figure 38. DEA measurements with two different processing compressions

The increase of the ionic viscosity is in correspondence with the slight exothermic peak of the temperature which indicates the cure of the material. After 200 seconds, the ionic viscosity shows no more significant changes which determines the end of the curing progress. The ionic viscosities with 80 bar and 120 bar show a similar curve progression. Therefore, an impact of the processing pressure on the curing behaviour was not proven. In general, by using the DEA method, the development of a cure kinetic model can be implemented with non-isothermal real-time measurements. Then, different heating rates have to be used [116]. However, more experience has been gained for the development of cure kinetics by using the DSC in the past. Therefore, the method by using the DSC was chosen to develop the cure kinetics in this study. The DSC measures exothermic and endothermic heat flow with a special laboratory test equipment. The heat flow changes in accordance to the curing progress. The rate of heat flow is proportional to the chemical cross-linking reactions by the thermoset [127]:

$$\frac{dH}{dt} = \Delta H_R \frac{d\alpha}{dt} \quad . \tag{4.4}$$

The cure kinetics are determined by the implementation of non-isothermal DSC measurements. Here, different heating rates are used (2.5, 5.0, 10.0, 15.0, 25.0, 50.0 K/min) with the DSC unit 'DSC 250 discovery' by TA Instruments. Each heating rate has been repeated three times to consider a mean. The tests have started from a temperature of 20 °C to 250 °C. No further exothermic effects were expected above 250 °C.

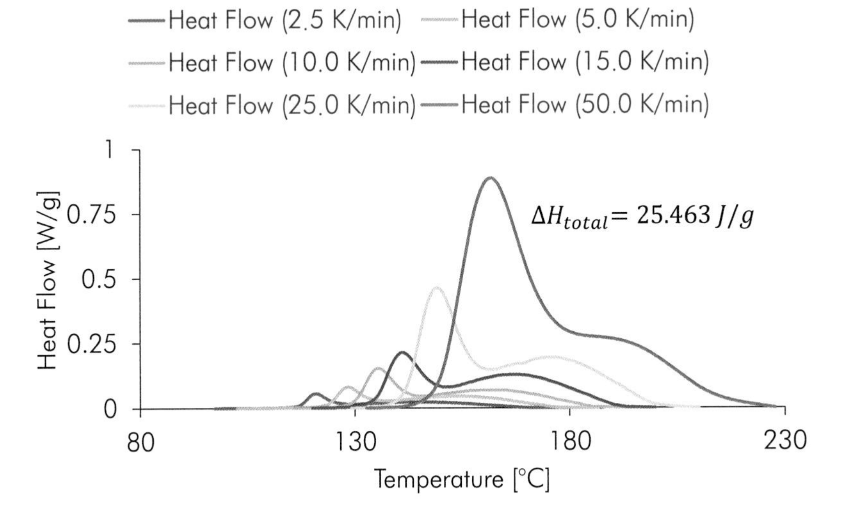

Figure 39. Non-isothermal DSC measurements using six different heating rates

Following statements can be made regarding the DSC test results:

1. Higher heating rates lead to higher exothermic peaks.
2. Higher heating rates lead to a delayed start of the exothermic reactions including the peak maximum.
3. The normalized heat flows show weak signals: a heating rate of 2.5 K/min leads to a maximum of 0.056 W/g. Even a heating rate of 50 K/min just leads to a maximum of 0.885 W/g.
4. The total heat flow of the SMC material is 25.463 J/g. Each DSC measurement was considered to calculate the average with a standard deviation of 13.4 %.
5. Two appearing peaks are in evidence: The first one is superior with a stronger increase in the beginning. The second follows on the first one which is flatter.

The comparison with the pure UP-resin and styrene makes the weakness of the heat flow signal clear. This conclusion was drawn by non-isothermal DSC measurements with the pure UP-resin and a heating rate of 10.0 K/min (Figure 40). The total heat flow of the pure UP-resin is 210.880 J/g. Therefore, it is more than eight times higher in comparison to the total heat flow of the SMC material. It seems that the additional filler components strongly inhibit the heat flow signal.

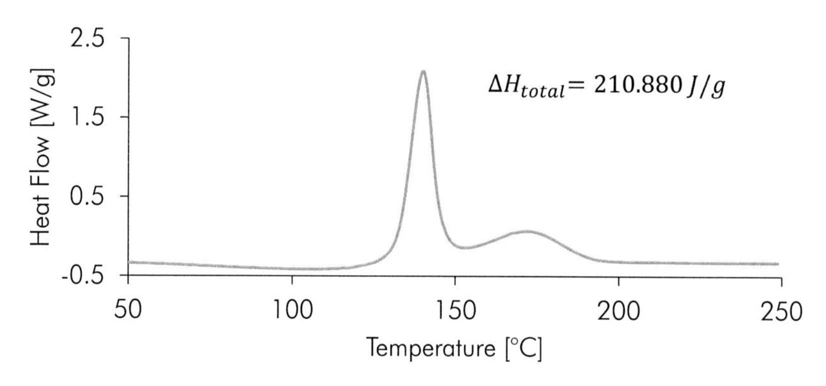

Figure 40. Non-isothermal DSC measurement (10 K/min) with the pure resin without any additives

In accordance to the DSC measurements of the SMC material, the pure UP-resin shows two subsequent heat flow signals (Figure 40). But especially the first peak is higher in comparison to the DSC measurements of the SMC material. The exothermic heat flow is not symmetrical pronounced for the SMC material as well as the UP-resin which could be caused by the existence of two independent exothermic reactions [127], [128]. A consecutive reaction is assumed similar to a series reaction:

$$A \rightarrow B \rightarrow C.$$

Reaction B-C is able to start, when the reaction A-B is, at least, partially completed. The first peak may be associated with the copolymerization initiated by peroxide decomposition [129]. The second peak is related to the thermally initially copolymerization. An increase of the peroxide amount produces a corresponding increase in the fraction of area of the first peak [129].
In the next step, the gelation temperature, which was determined by using the non-isothermal rheology measurements, was transferred to the gelation point at the degree of cure. Figure 41 shows the non-isothermal DSC measurement at a heating temperature of 2.5 K/min. In addition, the gelation temperature, which was determined for a heating rate of 2.5 K/min, is marked.

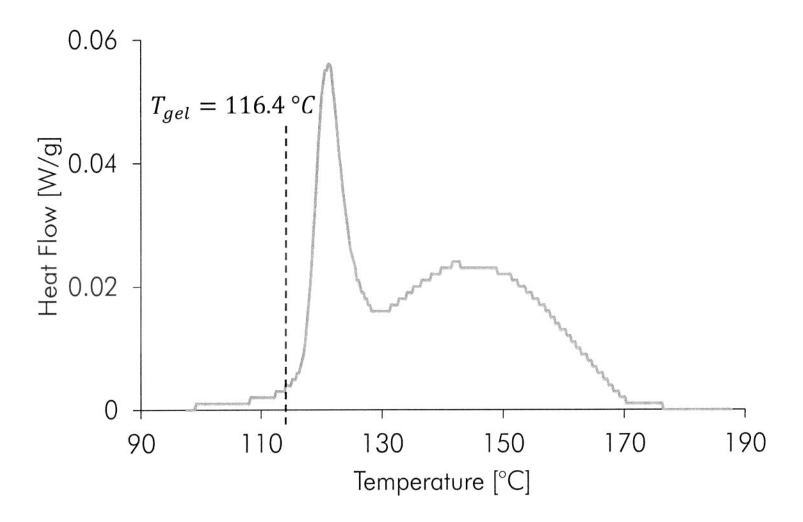

Figure 41. Non-isothermal DSC measurement with a heating rate of 2.5 K/min

The ratio of the heat flow signal at the gelation point and the overall material heat flow signal determines the degree of cure at the gelation point (Equation 4.4). This is illustrated in Figure 42 which shows the transformation of the heat flow signal into the curing progress. In addition, the loss angle is shown with the criterion for the determination of the gelation point ($\tan \delta = 1$). The gelation point was found at a degree of cure of 2 %.

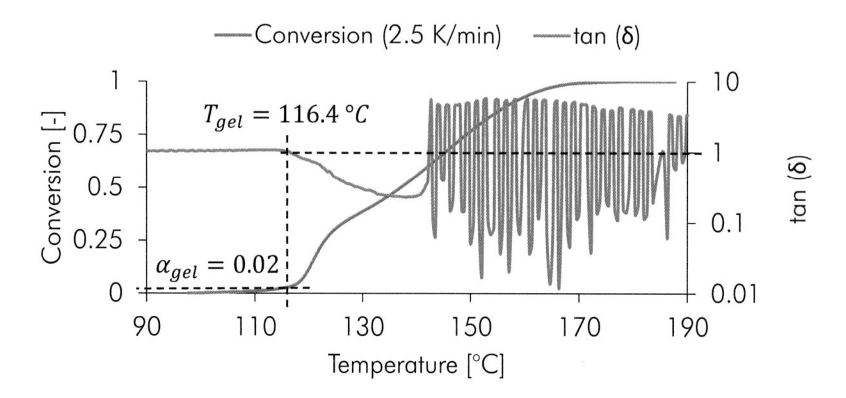

Figure 42. Determination of the gelation point

The determination of the gelation point can be analogical realized with each heating rate. Figure 42 considers a heating of 2.5 K/min because this heating rate was exemplary used to illustrate the determination of the gelation temperature (Figure 36).
After the determination of the gelation point, the cure kinetic model can be developed. Taking the expanded Prout-Tompkins-Equation and the two subsequent chemical reactions into account, the final differential equation can be expressed as the following:

$$\frac{d\alpha}{dt} = C_1 k_1 \alpha^{m_1}(1-\alpha)^{n_1} + C_2 k_2 \alpha^{m_2}(1-\alpha)^{n_2} \quad . \tag{4.5}$$

The constants k_1 and k_2 describe the Arrhenius-Temperature with the empirical constants A_1, A_2, E_1 and E_2:

$$k_1 = A_1 e^{-\frac{E_1}{RT_K}} \quad , \tag{4.6}$$

$$k_2 = A_2 e^{-\frac{E_2}{RT_K}} \quad . \tag{4.7}$$

The empirical parameters C_1, C_2, A_1, A_2, E_1, E_2, m_1, m_2, n_1, and n_2 were fitted to the results of the non-isothermal DSC measurements. It is important that the cure kinetic model illustrates the period up from the beginning to the gelation point with the highest degree of accuracy. The empirical parameters are summarized in Table 6.
After the determination of the empirical parameters, the differential equation works for each (isothermal) curing temperature to determine the curing progress with respect to time. Therefore, the time between the beginning of the process and time of passing the gelation point can be calculated.

Table 7 summarizes the results for temperatures between 110 and 150 °C.

Table 6. Developed parameters for the expanded Prout-Tompkins model

Parameter	Unit	Reaction 1 (A-B)	Reaction 2 (B-C)
$[C]$	1	0.448	0.552
$[A]$	1/s	9.12E+11	3.57E+14
$[E]$	J/mol	102815	137104
$[n]$	1	1.637	1.280
$[m]$	1	0.802	0.001

Table 7. Gelation time in dependence on different curing temperatures

Temperature [°C]	110	120	130	140	150
Gelation time $[t_{gel}]$	309.3	136.6	63.0	30.5	15.4

Figure 43 illustrates the progress of the degree of cure by using isothermal temperatures. An exponential relation of the curing temperature and the gelation point can be observed (Figure 44).

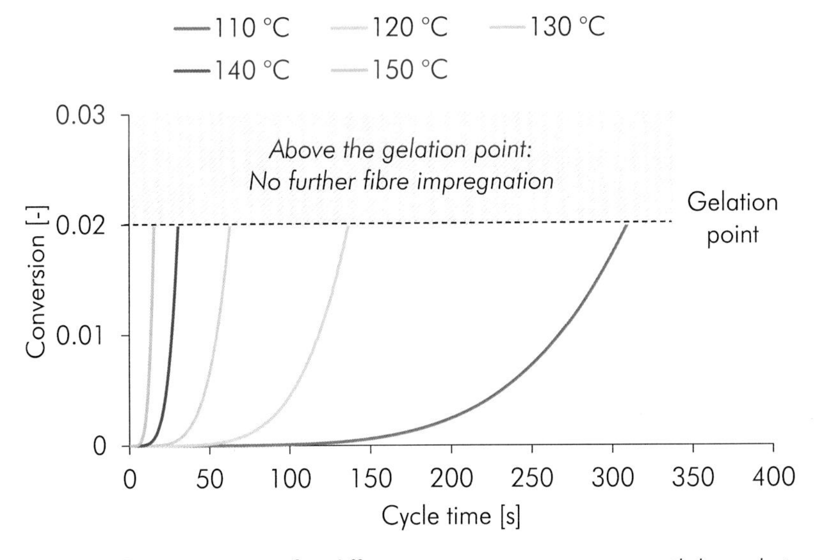

Figure 43. Curing progress for different curing temperatures until the gelation point is passed

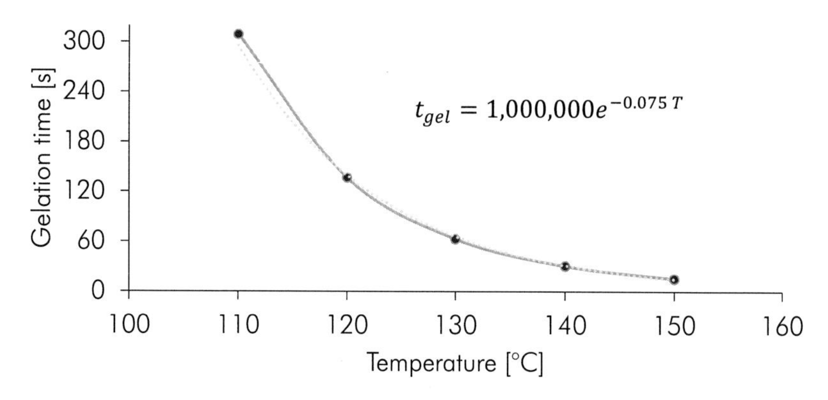

Figure 44. Temperature-gelation point dependence of the SMC material

4.1.3 Viscosity

Chemo-rheology describes the change of the viscosity due to the chemical reaction of polymers. Ryan has defined that the chemo-viscosity of thermosets is a function of the pressure gradient, the temperature, the time, the shear rate, and the filler properties [130]:

$$\eta = \eta(T, p, \dot{\gamma}, t, F) \quad . \tag{4.8}$$

However, there are no analytical models available which consider all of these parameters. Most of the analytical models just consider one or two parameters. Cure effects η_c, shear rate effects η_{sr} and filler effects η_f are tested in individual tests for the development of a chemo-rheological model:

$$\eta_c = \eta_c(T, t) \quad , \tag{4.9}$$

$$\eta_{sr} = \eta_{sr}(T, \dot{\gamma}) \quad , \tag{4.10}$$

$$\eta_f = \eta_f(F) \quad . \tag{4.11}$$

In general, SMC materials based on UP-resins are non-Newtonian, shear thinning, and viscoelastic [131]. In the past, no dependency of the shear rate on the viscosity has been observed for unsaturated polyester resins before the gelation point. Therefore, shear rate effects on the fibre impregnation are neglected [132]. Furthermore, filler effects are not considered because the exact composition of the filler materials within HUP63 is not known and no modifications are implemented. Therefore, the final viscosity model will depend on the temperature with respect to time.

Non-isothermal viscosity measurements are done to determine the general viscosity trend of the SMC material HUP63. The tests are implemented by using the oscillated rheometer unit AR2000 by TA instruments. Figure 45 shows the complex viscosity using a heating rate of 5 K/min. Three different frequencies were used.

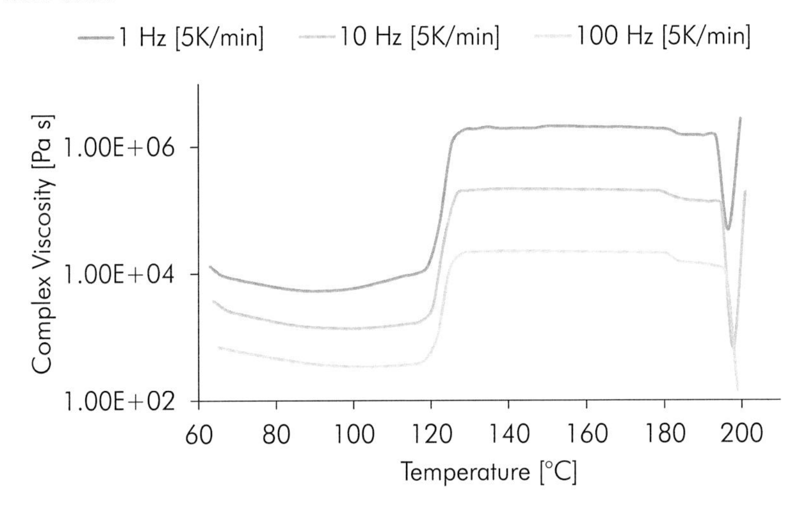

Figure 45. Non-isothermal rheological measurements with a heating rate of 5 K/min and three frequencies

In general, lower frequencies lead to a higher complex viscosity but the frequency does not change the general viscosity behaviour. Before passing the temperature of 120 °C, the complex viscosity slightly decreases. A fast increase of the complex viscosity can be seen at a temperature of 120 °C up to a complex viscosity with a factor of 100. Then, the complex viscosity remains constant until a temperature of 195 °C. Already the non-isothermal viscosity measurements with a low heating rate shows the sudden increase of the viscosity.
The compression moulding of Hybrid SMC composites is an isothermal process that is why isothermal viscosity measurements were implemented by using the same the oscillated rheometer unit. Different isothermal temperatures were chosen between 90 and 150 °C for the rheometer measurements. Depending on the specific gelation temperatures, the isothermal temperatures were kept between 2 and 20 minutes. A frequency of 5 Hz and a deformation rate of 1 % were chosen. Figure 46 shows the time- and temperature-dependent viscosity behaviour until the gelation point is reached. It can be seen that the material's viscosity rises with higher curing temperatures. The initial viscosity starts between 1,000 and 2,500 Pa s independent on the test temperature.

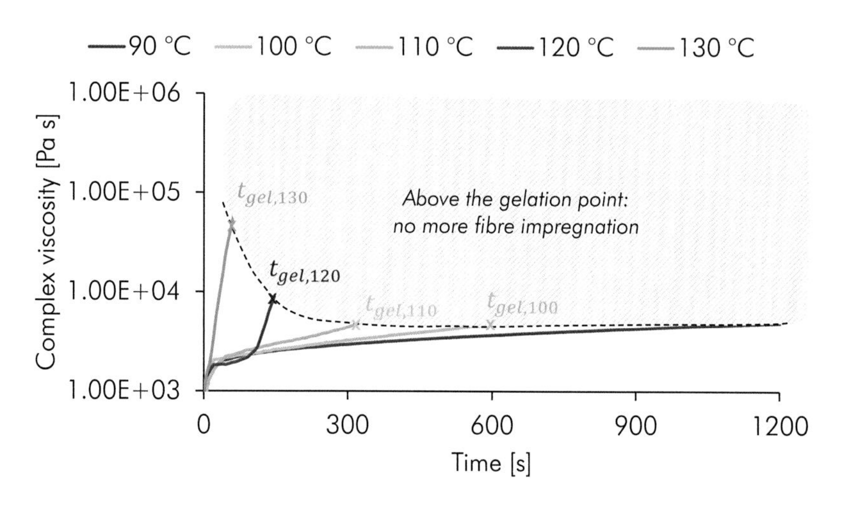

Figure 46. Isothermal viscosity measurements of the SMC material HUP63

A time- and temperature-dependent viscosity model has to be developed based on the results of Figure 46. In the past, the Castro-Macosko model was often used to describe the progress of the viscosity during processing. It already considers the gelation point to determine the temperature-dependent viscosity [133], [134]:

$$\frac{\eta(T_K)}{\eta_0(T_K)} = \left(\frac{\alpha_{gel}}{\alpha_{gel} - \alpha}\right)^{CM_1+\alpha CM_2} \quad \text{with} \tag{4.12}$$

$$\eta_0(T_K) = A_{CM} exp\left(\frac{T_b}{T_K}\right) \quad . \tag{4.13}$$

A_{CM}, CM_1, CM_2 and T_b are empirical parameters. In the past, the Castro-Macosko model has been already used even for SMC materials [135], for example considering in-mould coating (IMC) [136]. However, it has not been used for the impregnation of a continuous carbon fibre reinforcement yet. For the viscosity of the SMC material HUP63, the empirical parameters could not be fitted in an appropriate way that is why other solutions have to be found. Therefore, an alternative analytical calculation has to be developed which represents the viscosity characteristics of HUP63. Each viscosity progress can be described with an exponential growth function with the basis 10 which includes the empirical parameters m and n:

$$\eta(t) = 10^{mt+n} \quad . \tag{4.14}$$

Each viscosity measurement is fitted with the least square method to determine and to optimize the empirical parameters m and n. The parameters are summarized in Table 8. To achieve a better illustration, each temperature is shown in an individual diagram because of the differences in time until the gelation point (Figure 47 - Figure 51). The result is a good fitting for each temperature. In the beginning of the measurements, slight deviations can be noticed which can be a result of the isothermal measurements by the rheometer and the accompanying temperature deviations in the beginning. Just the complex viscosity at a temperature of 120 °C do not follow the general trend because the measurements look similar to a S-shape which differs to the exponential growth function (Figure 50). Despite several repetitions, the measurements at a temperature of 120 °C may indicate an outlier.

Table 8. Fitted parameters for the analytical function

Isothermal temperature	m	n
90 °C	0.00022	3.41176
100 °C	0.00062	3.30583
110 °C	0.00136	3.24393
120 °C	0.00539	2.96644
130 °C	0.02556	3.19227

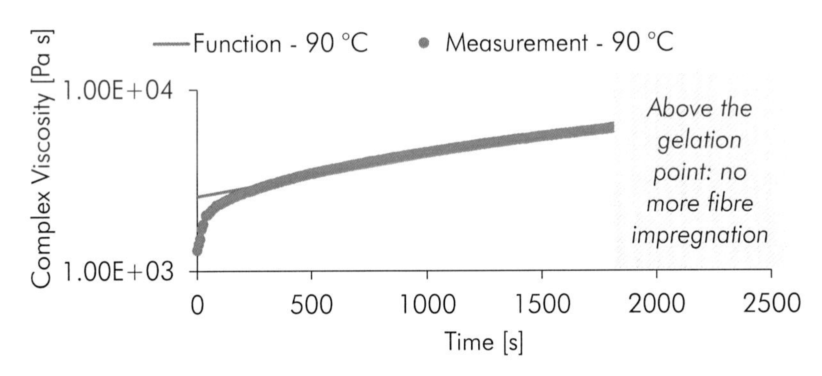

Figure 47. Isothermal viscosity measurements with 90 °C and fitted function

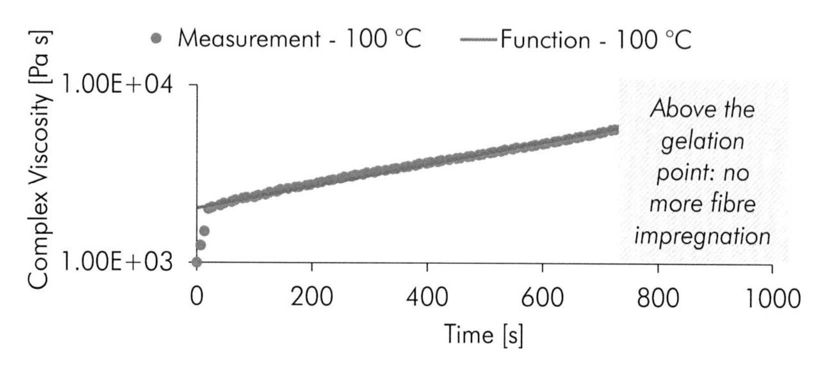

Figure 48. Isothermal viscosity measurements with 100 °C and fitted function

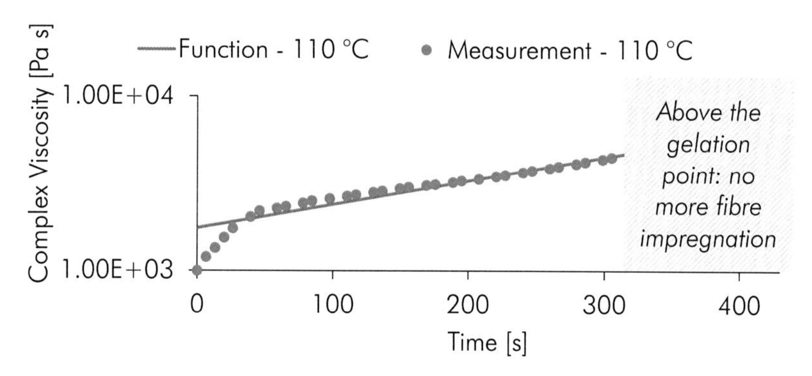

Figure 49. Isothermal viscosity measurements with 110 °C and fitted function

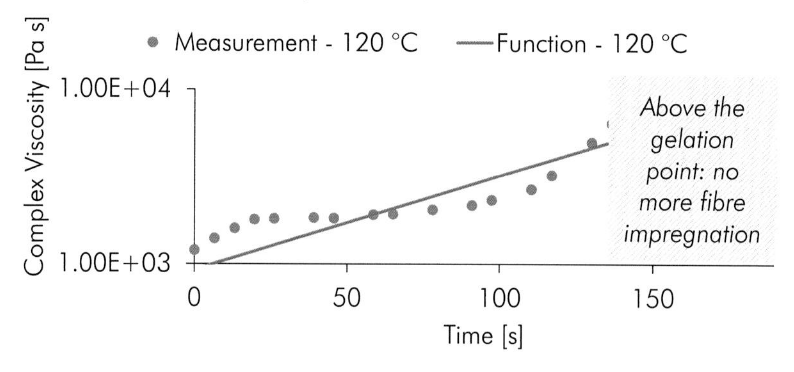

Figure 50. Isothermal viscosity measurements with 120 °C and fitted function

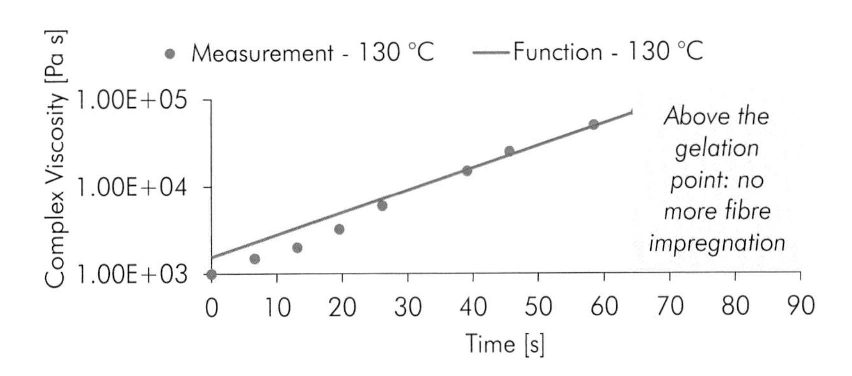

Figure 51. Isothermal viscosity measurements with 130 °C and fitted function

The individual values for the empirical parameters m and n are transferred to a graph (Figure 52). Isothermal viscosity measurements were unable to use for the temperatures above 140 °C because of the fast curing reactions. Therefore, a trend analysis was implemented for the estimation of the m- and n-values. In addition, polynomic functions of the third grade are able to describe the empirical parameter m and n in an appropriate way:

$$m = 6.7 \cdot 10^{-7} T^3 - 0.00019\, T^2 + 0{,}0179\, T + 0.5566 \quad , \qquad (4.15)$$

$$n = -2.1 \cdot 10^{-6} T^3 + 0.00082\, T^2 - 0.1107\, T + 8.2165 \quad . \qquad (4.16)$$

Note, that the polynomic functions for the analysis of n does not consider the n-value at a measurement of 120 °C. It is handled as an outlier. The coefficient of determination is high for both functions ($R_m^2 = 0.9980$ and $R_n^2 = 0.9995$).

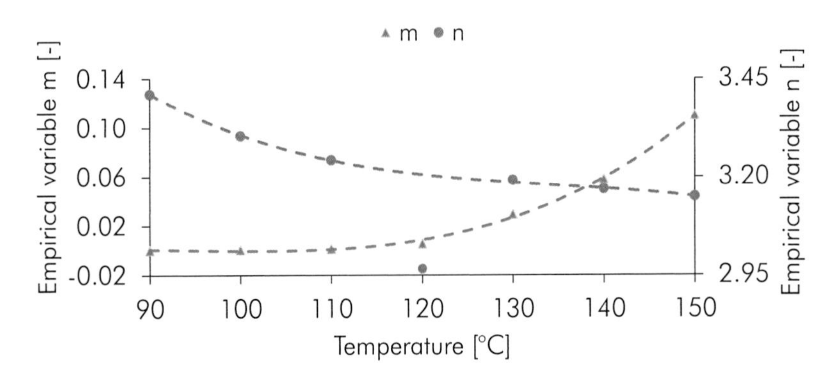

Figure 52. Empirical parameters m and n in dependence on the temperature

The m-value becomes negative between a temperature of 96 °C and 107 °C which strongly influences the analytical description of the viscosity. An investigation of the impact is necessary while the development of the impregnation model. The transfer of the parameters leads to a the predicted viscosity development for temperatures of 140 °C and 150 °C until the gelation point (Figure 53). Because of reasons of clarity, just the temperatures between 120 °C to 150 °C are illustrated in Figure 53.

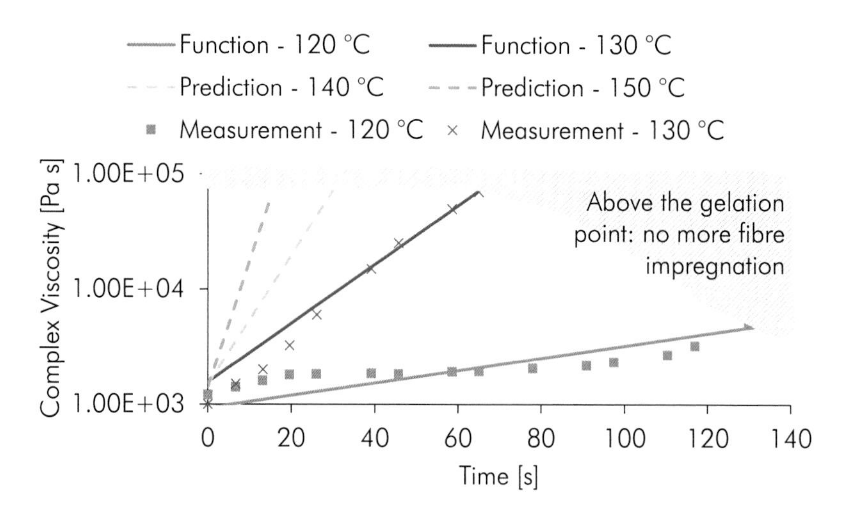

Figure 53. Isothermal viscosity measurements between 120 °C and 150 °C

4.2 Processing compression

The processing compression is the initiating and the driving force to transfer the resin from the SMC material between the individual carbon fibre filaments. During the overall process, the processing compression is isobaric, so it does not change during the compression moulding. However, there is a pressure gradient which changes along the impregnation length. During the fibre impregnation of the continuous carbon fibre reinforcement by the SMC material, the pressure decreases with the advancing flow front. In the end of the fibre impregnation, the pressure is decreased to zero. Because of the simultaneous fibre impregnation of the carbon fibre reinforcement by the upper and lower side, the reinforcement structure can be subdivided into two symmetrical parts with the same pressure gradient (Figure 54).

Therefore, the pressure gradient p_{fluid} is a function of the applied processing compression p_{app}:

$$p_{fluid} = f(p_{app}) \quad . \tag{4.17}$$

Despite the isobaric process conditions, there are further effects on the processing compression. First, there is the readjustment of the hot press. It takes place when the material shrinks, expands or cures. The readjustment differs from the adjusted setting of the hot press as well as from the specific material. Second, in the beginning of the process, the processing compression raises from zero to the desired pressure. The raise is dependent on the closing speed of the mould. Greater closing speeds lead to a faster increase of the processing compression.

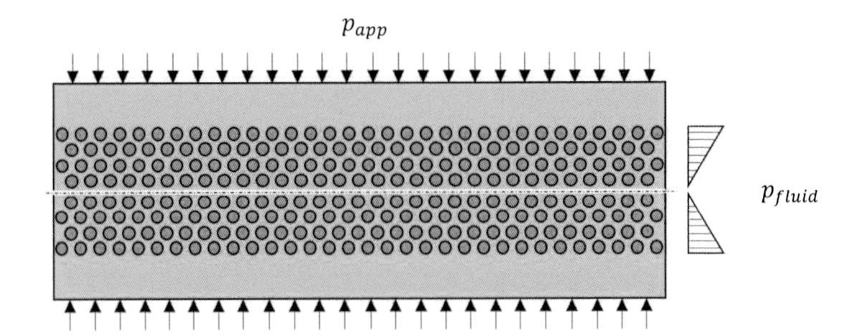

Figure 54. Pressure gradient during the manufacturing of Hybrid SMC composites

4.3 Textile parameters

In this study, the focus is on unidirectionally arranged continuous carbon fibre reinforcements in the form of a TFP textile. Especially, two characteristics are essential for the analysis of the fibre impregnation by a SMC material. On the one hand, there is the textile's permeability which is mainly driven by the fibre volume content and the fibre diameter. On the other hand, there is the thickness of the reinforcing textile which determines the impregnation length. Both parameters are investigated in the following.

4.3.1 Permeability

The permeability describes the directional impregnation of a semi-finished textile. It is described by the anisotropic tensor of second rank [137].

$$K = \begin{bmatrix} K_{xx} & K_{xy} & K_{xz} \\ K_{yx} & K_{yy} & K_{yz} \\ K_{zx} & K_{zy} & K_{zz} \end{bmatrix} = \begin{bmatrix} K_1 & 0 & 0 \\ 0 & K_2 & 0 \\ 0 & 0 & K_3 \end{bmatrix} . \quad (4.18)$$

K_1 is the permeability longitudinal and K_2 is the permeability perpendicular to the fibre direction. The permeability factor K_3 represents the permeability through the thickness of the semi-finished textile.
The manufacturing of Hybrid SMC composites is characterized by a core layer of unidirectionally arranged carbon fibre reinforcements which is covered by two layers of SMC material. A mould coverage of 95 % leads to an almost exclusive fibre impregnation through the thickness of the textile. Therefore, the permeability reduces to a one-dimensional case.

$$K = K_{zz} = \begin{bmatrix} 0 & 0 & 0 \\ 0 & 0 & 0 \\ 0 & 0 & K_3 \end{bmatrix} . \quad (4.19)$$

The textile's permeability can be experimentally determined. However, the experimental determination of the permeability of TFP textiles offers difficulties. Usually, TFP textiles are designed as individually adapted reinforcing structures which do not represent a uniform textile design in comparison to conventional semi-finished textiles. Furthermore, there are further individual parameters by using the TFP technology which increase the complexity in the terms of permeability. Maybe these are reasons why the state of art still has a lack of knowledge regarding the permeability of TFP textiles. Therefore, analytical models are used in this study to describe the permeability through the thickness of a reinforcing textile. In the past, several analytical models have been developed. Advani and Sozer have given a brief overview about the most common analytical models [138]. The permeability models of Bruschke and Advani [139], Gutowski et al. [140] and Gebart [141] are illustrated in Figure 55. All analytical models are dependent on the radius of the fibre reinforcement ($r_{cf} = 3.5\ \mu m$) and the fibre volume content.
The model of Gutowski et al. works with empirical parameters. They have given recommendations for the empirical parameters for the impregnation through the thickness which have proven their value in the past [142]. Therefore, Figure 55 describes two models of Gutowski et al. which both consider the recommended parameters with their minimum (K_Gut_min) and maximum values (K_Gut_max). Gebart has made a difference between a quadratic (K_Geb_quad) and hexagonal fibre arrangements (K_Geb_hex) that is why two models by Gebart are considered. The models show a high accuracy to each other for fibre volume contents between 50 and 60 %. With an increasing fibre volume content, the calculated permeabilities show higher deviations to each other. At lower fibre volume contents, the model of Bruschke and Advani is

almost equal to the model of Gebart with a quadratic fibre arrangement but even the consideration of a hexagonal fibre arrangement is close to these results. The Gutowski model shows lower permeabilities in comparison to the other models.

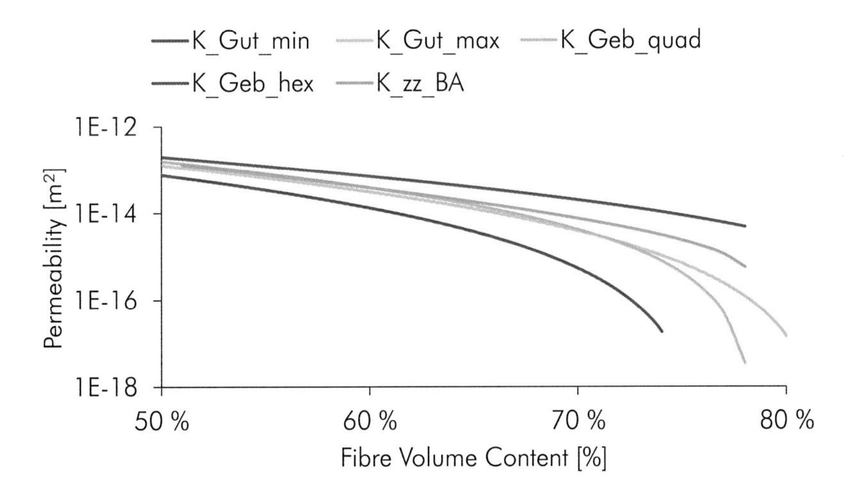

Figure 55. Analytical models for the determination of the permeability

Actually, it is difficult to evaluate the analytical models for TFP reinforcements within the Hybrid SMC compression moulding from the current point of view because of some reasons. First, beside the existence of the carbon fibre reinforcement, the woven fabric as a base material significantly influences the permeability. Second, misalignments of individual fibres or plies influences the permeability as well because fibre misalignments create alternative flow paths for the resin [143]. It was already demonstrated that the stitch density has an impact on fibre misalignments (Figure 23). Third, the sewing thread is not considered in the presented analytical models. However, Drapier et al. have proven in their studies that an increase of the stitch density leads to a proportional increase of the permeability through the thickness [144]. The resistance of the fluid's flow is reduced to the increased stitches which are used by the resin to flow along the sewing thread. This was proven for Hybrid SMC composites with the implementation of one-sided fibre impregnation tests. The impregnation of a continuous carbon fibre reinforcements in combination with the base material was performed by a SMC material which was put on the top of the fibre reinforcement. The non-covered side with the base material was microscopically investigated (Figure 56).

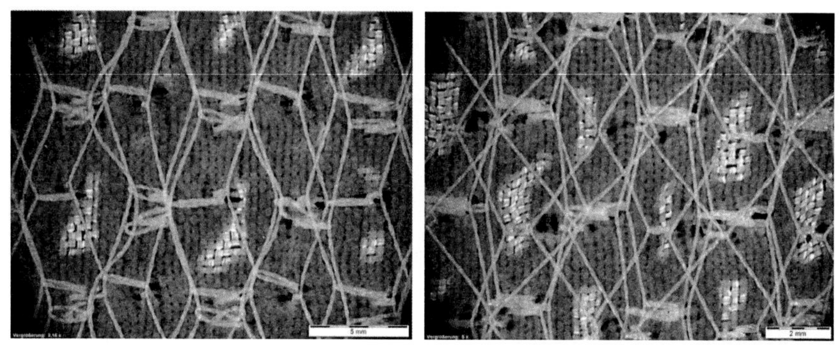

Figure 56. Dry spots at areas without sewing thread

The left image is taken at a TFP textile with a lower thickness of the carbon fibre reinforcement in contrast to the right image. Due to the lower thickness of the reinforcement, a lower number of sewing threads can be noticed. Most of the dry areas are located at the base material which was not treated with a sewing thread. Therefore, the formation of flow channels at the sewing thread can be assumed.

4.3.2 Thickness

The thickness of the fibre reinforcement determines the required flow path and impregnation length, respectively. In the beginning of the compression moulding process, a consistent thickness does not exist, because of the flexible textile properties. A cut-out of the unidirectional carbon fibre textile, which is used in this study, is shown in Figure 57. The mapping of the surface was taken with the digital 4K-microscope VHX-7000 by Keyence. Already this cut-out shows thickness deviations of the fibre reinforcement of more than 0.3 mm.

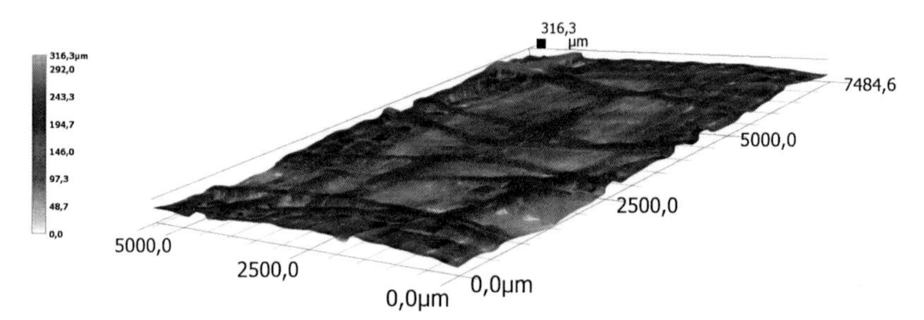

Figure 57. Thickness deviations of a non-impregnated semi-finished textile

However, the processing compression during the compression moulding makes the reinforcing textile flat and uniform. In a previous study by Duhovic et al.

[109], the following assumption was defined. After the compression moulding starts, the first step is characterized by the compaction of the fibre reinforcement. Only when this step is accomplished, the fibre reinforcement is able to become impregnated. Therefore, the total impregnation length is assumed to be constant:

$$\Delta z = const \quad . \tag{4.20}$$

The quantification of the thickness was evaluated in an experimental study. A stacking sequence was chosen which represents a TFP carbon fibre reinforcement in the core which is covered by two layers of SMC material. By considering the second TFP design (compare Figure 25), a bundle distance d_{fb} of 4 mm was chosen. When a smaller fibre bundle distance was used, it refers to the third TFP design (compare Figure 25). Different thickness levels were implemented by using different amount of carbon fibre layers n_L on the glass fibre base material. A temperature of 135 °C and a processing compression of 120 bar was chosen. Note that the reinforcement's thickness does not change with other parameters. The thickness was studied by image analysis similar to the methods (chapter 3.3.2). The average thickness of five Hybrid SMC laminates were investigated to consider the standard deviations. In addition, the thickness for a fibre bundle distance of 2 mm and 3 mm were calculated by considering the fibre volume content because the fibre volume content does not change and therefore does not influence the thickness of the reinforcement. A fibre volume content of 58 % was found for these tests. Note that the fibre volume content has shown no dependency on the process conditions. The results of the measurements and the additional calculations are shown in Figure 58.

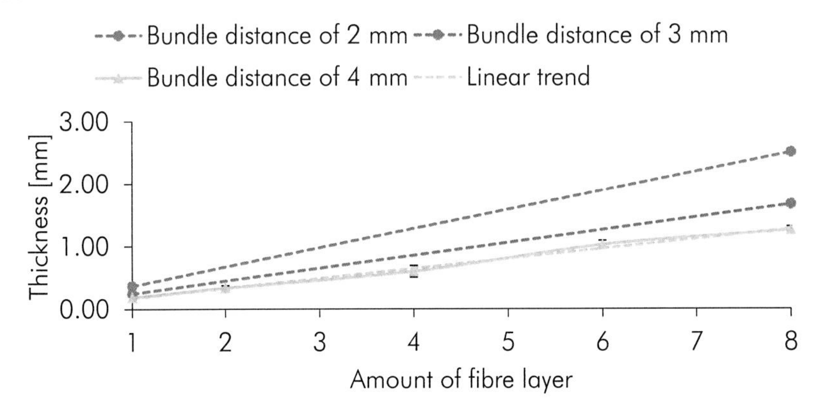

Figure 58. Change of thickness in dependence on the amount of fibre layer

With an decreasing fibre bundle distance d_{fb}, the total thickness z_{total} increases. The thickness of the base material z_{bm} has to be added to the overall calculation of the continuous carbon fibre reinforcement. The base material's thickness z_{bm} in this study is 0.095 mm. The total thickness can be calculated with a linear equation which consider the set fibre bundle distance d_{fb} and the total amount of carbon fibre layer n_L:

$$z_{total} = z_{CF} + z_{bm} = 0.64\,\frac{n_L}{d_{fb}} + 0.095 \quad . \tag{4.21}$$

The test specimens are characterized by a low standard deviation. Hybrid SMC composites with six or eight layers have shown an incomplete fibre impregnation in the middle of the reinforcing carbon fibre structure (Figure 59) but the linear trend still fit to the total thickness. Non-impregnated areas within the fibre reinforcement are experimentally investigated in the following chapter.

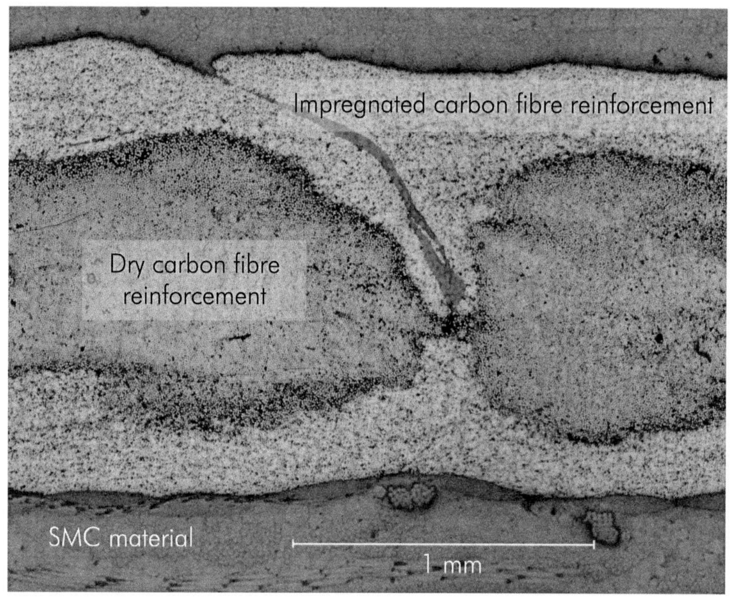

Figure 59. Poor impregnation of a CF reinforcement with eight layers

5 Experiments

This chapter deals with the experimental investigations to determine the impact of the textile and process parameters on the fibre impregnation within Hybrid SMC composites. The experiments are divided into three parts. The first part describes the experimental tests of load transferring architectures which support the analytical impregnation model as a validation instrument. The second part is about the load introducing fibre architecture. The additional parameter of load introducing fibre architectures are qualitatively evaluated in individual experiments. The third part deals with the presentation of an additional process step before compression moulding which contains the pre-impregnation of the TFP textiles by using an impregnation lane for SMC materials. Here, the degree of impregnation was evaluated and the advantages as well as disadvantages are presented.

5.1 Load Transferring Fibre Architecture

Load transferring fibre architectures are straight fibre geometries which guide loads from load introducing areas into the composite structure. Exclusive unidirectionally arranged fibre geometries were implemented in the experimental tests to systematically investigate the fibre impregnation. The reinforcing textile is as large as the size of the mould (250 x 120 mm). When the end of the unidirectional arrangement of the carbon fibre bundle is reached, it changes direction by 180 degrees, so that each carbon fibre bundle lays next to each other considering a fibre bundle distance d_{fb} (Figure 60).

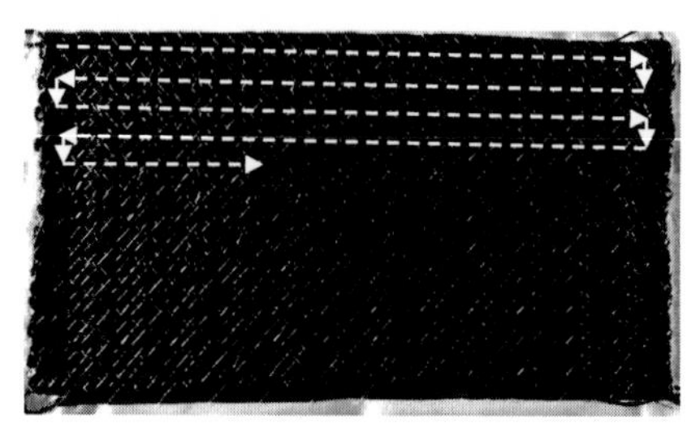

Figure 60. TFP textile with unidirectionally arranged carbon fibre bundles

5.1.1 Textile and Process Parameters

Several factors can influence the carbon fibre impregnation by the SMC material during the compression moulding process. These factors can be classified into the material characteristics and the process conditions. A further specification can be made on the materials which can be subdivided into the individual textile and SMC characteristics. The material characteristics, which can have an impact on the fibre impregnation, are summarized in the following table.

Table 9. Material parameters which can affect the fibre impregnation

Material characteristics	
Textile characteristics	SMC characteristics
Type of carbon fibre	Glass fibre content
Type of base material	Type of fibres
Type of sewing thread	Resin content
Fibre bundle distance	Surface weight
Amount of carbon fibre layers	Viscosity
Stitch formation	Gelation point
Stitch density	Filler content
	Type of fillers
	Mould coverage
	Storage time

The compression moulding of SMC is an isothermal and isobaric process. The individual process parameters can have an impact on the fibre impregnation. The parameters are summarized in Table 10. The insertion time represents the period between the material is placed into the mould and the compression moulding begins. Thus, the SMC material heats up by the isothermal mould before the mould is completely closed and started.

Table 10. Process parameters which can affect the fibre impregnation

Process parameters
Temperature
Processing compression
Time
Insertion time
Mould closing velocity

All in all, many types of parameters were presented which can influence the fibre impregnation by SMC materials. The evaluation of each parameter's impact requires an extraordinary scope regarding time and costs. Therefore, the most important parameters are chosen to reduce the parameters within an experimental design. Other parameters are not considered in the experimental study. The SMC material HUP63 was chosen for this study. It is not pursued to modify the SMC material that is why all parameters, which belong to the SMC material, can be neglected. In addition, a mould coverage of 95 % is always used and leads to the exclusive fibre impregnation through the thickness. The individual fibre materials within the TFP textile is not changed as well. Furthermore, the TFP textiles are always processed with the same stitch formation. Regarding the process parameters, a variation of the temperature and a variation of the processing compression is considered. Other parameters are kept constant. Figure 61 shows an overview of the individual parameters which can affect the fibre impregnation.

Influencing factors:
1. Processing compression
2. Stitch density
3. Thickness of the reinforcement
4. Temperature

Not considered factors:
1. Alternative Materials
2. Mould coverage
3. Ply arrangement
4. Velocity of mould closing

Disturbance variable:
1. Heating of SMC plies
2. Storage time of SMC material

Figure 61. Influencing parameters in the experimental tests

In accordance to Siebertz et al., the parameters were subdivided into influencing factors, not considered factors, and disturbance variables [145]. Only the influencing factors are considered in the experimental design. Finally, four parameters were identified to investigate their influence on the fibre impregnation by SMC materials. Two of the four parameters belong to the compression moulding process. The stitch density as well as the reinforcement's thickness belong to the properties of the TFP textile. In contrast to the process parameters, the textile parameters require an analytical calculation. In the case of Hybrid SMC composites, the stitch density is correlated with the number of flow channels through the carbon fibre reinforcement. The stitch density is defined as the amount of stitches through the thickness per square millimetre. The fibre bundle distance d_{fb} and the stitch length d_{sl} determines the required area for each stitch. This represents the divisor within the calculation. In addition, taking the amount of layers n_L into account, the number of flow channels increases with the factor of the Gaussian Sum Formula. This is illustrated by the following example. The fixing sewing thread of the second layer additionally stitch through the first layer to fix it with the base material. The fixing sewing thread of the third carbon fibre layer stitch through the second and the first layer, and so on. Equation 5.1 presents the final calculation for the stitch density ϑ and the number of flow channels through the fibre reinforcement, respectively. This equation assumes a constant stitch density. In reality, the number of flow channels decrease with the increase distance to the base material. Therefore, the stitch density is dependent on the flow length during the fibre impregnation which is dependent on the time. When two TFP reinforcements are located in the core of the Hybrid SMC compression moulding process, the fibre impregnation of the first carbon fibre layer takes place first. Here, most of the flow channels are integrated according to the Gaussian Sum Formula. It represents the reality better than an exclusive factor of the amount of fibre layers or an exclusive description by the stitch length. Therefore, equation 5.1 is chosen for a calculation of the stitch density or the number of flow channels.
The analytical calculation for the thickness of the reinforcement was already presented in equation 4.21. Due to the symmetric stacking sequence, the half of the reinforcement's thickness have to be considered z_{half}. Here, the crucial variables are fibre bundle distance d_{fb} and the amount of carbon fibre layers n_L as well as the thickness of the base material z_{bm} (Equation 5.2):

Stitch density
$$\vartheta = \frac{\frac{n_L}{2}(n_L + 1)}{d_{fb}\, d_{sl}} \quad . \tag{5.1}$$

Reinforcement's Thickness
$$z_{half} = 0.64\,\frac{n_L}{d_{fb}} + 0.095 \quad . \tag{5.2}$$

Taking a closer look at the influencing and chosen parameters, a relation to the parameters of Darcy's Law can be found (Table 11).

Table 11. Parameter in Darcy's Law and range within the experimental tests

Experimental parameter	Variable	Darcy's Law	Minimum	Maximum
Stitch density	ϑ	K_z	0.045 st./mm^2	1.636 st./mm^2
Reinforcement's thickness	z_{half}	Δz	0.128 mm	1.328 mm
Processing compression	ΔP	ΔP	40 bar	160 bar
Temperature	T	η	110 °C	155 °C

The stitch density can be related to the permeability because it sets the number of flow channels through the thickness of the fibre reinforcement. The thickness of the textile describes the impregnation length. The pressure is driven by the hot press during compression moulding. The viscosity can be related to the temperature because temperature-dependent viscosity model was developed. Table 11 presents the individual parameters with their minimum as well as maximum values for the experiments. In the experiments, temperatures between 110 and 155 °C were chosen. Lower temperatures than 110 °C lead to long curing cycles which do not meet the requirements for an economic process. In contrast, higher temperatures than 155 °C lead to early stage curing reactions which endanger the fibre impregnation. The processing compressions were chosen with the help of the material data sheet (80 – 120 bar). In addition, lower and higher processing compression were added (40 and 160 bar). The calculation for the thickness of the reinforcement was chosen between 0.128 and 1.328 mm. Note that this thickness describes the half of the total reinforcement z_{half} because a double-sided impregnation takes place. The stitch density was chosen between 0.045 and 1.636 stitches per square millimetre. This value was calculated by equation 5.1. Note that this value is an average value with respect the amount of fibre layers n_L. Exemplary unidirectional TFP textiles are shown with a close-up view in Figure 62.
The four TFP textiles differ in the distance of the carbon fibre bundles to each other d_{fb} and the stitch length d_{sl}. Figure 62 shows bundle distances of one (A1) and two (A2) millimetres as well as the stitch lengths of ten (B10) and 25 millimetres (B25).

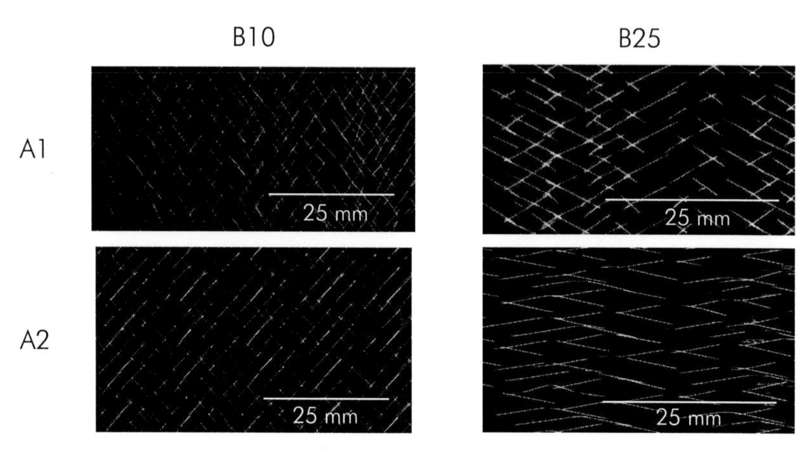

Figure 62. TFP textiles with different bundle distance and stitch density

5.1.2 Methodology and experimental procedure

The experimental design was developed with the principles of statistical design of experiments. However, classic two-factorial Design of Experiment (DoE) was not chosen for this investigation because more than two or three factor levels were investigated in this study. The individual parameters were randomly chosen and the experimental tests were sequentially implemented. The experimental tests can be subdivided into one-sided and double-sided fibre impregnation tests. The double-sided fibre impregnation is characterized by two SMC cover layers on both sides of the reinforcing textile. This leads to a reduction of flow paths and two converging flow fronts. This experimental setup was mainly used. The one-sided fibre impregnation tests just consist of one SMC cover layer which impregnates the carbon fibre reinforcement.

At first, Hybrid SMC composites with the specific material characteristics and process parameters were manufactured. The cycle time was individually chosen in dependence on the curing temperature but at least above the temperature- and time-dependent gelation point. After the manufacturing of the Hybrid SMC laminates, the specimens were prepared for the void analysis. Due to the high amount of Hybrid SMC laminates within the DoE, a fast, economic, and appropriate evaluation method was required. Here, microscopic image evaluations of polished cross sections of the Hybrid SMC laminates were chosen as major investigation technique. Three cut-outs of each Hybrid SMC laminate were prepared for the polished cross sections (Figure 63). The cut-out was always prepared perpendicular to the fibre direction because it offers a clear evaluation for the fibre impregnation between each single carbon fibre filament. A length of 40 millimetres was chosen for the cut-outs. The validation of light microscopy technique was realized by x-ray microscope scans of single specimens.

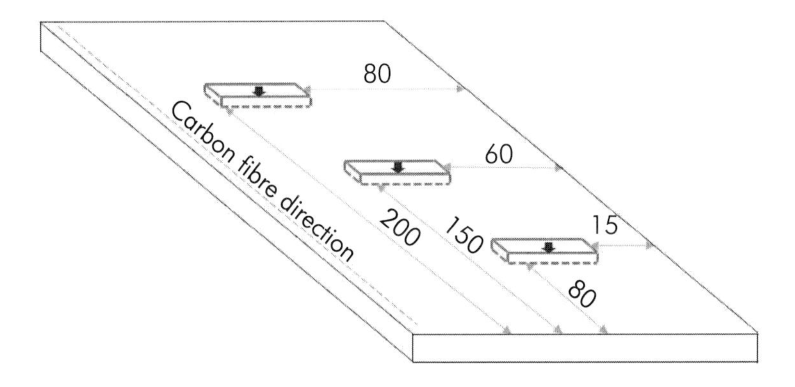

Figure 63. Cut-out positions of the Hybrid SMC composites for the preparation of polished cross-sections

The individual cut-outs were prepared for cross-sections by grinding and polishing. The polished cross-sections were analysed by a light microscope 'Axioplan' supplied by Zeiss. Most of the time, a 50 times magnification was used to capture the cross-section area. Image analysis was used to evaluate the FVC as well as the void content. The first step within the image analysis was the selection of the area with the carbon fibre reinforcement. After that, a threshold analysis was applied by setting the colour controller between the grayscale of the carbon fibres and the greyscale of the resin. The transition area between the voids and the resin has to be individually set (Figure 64) but it can be kept constant when the properties of the light microscope were kept constant as well.

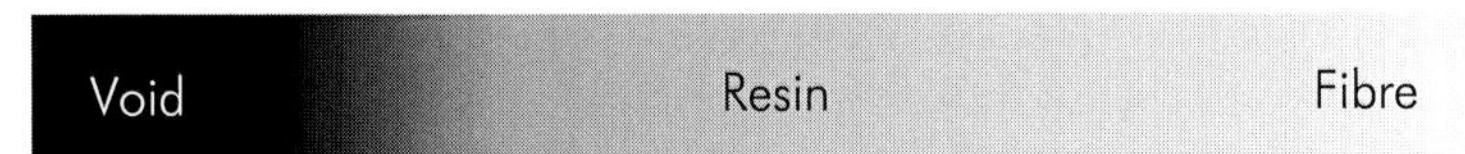

Figure 64. Grey scale and transition area while threshold analysis

A histogram analysis gives information about the percentage of the colour parts. By setting the controller between the grey scale of the resin and the grey scale of the voids, the void content within the reinforcement area can be determined with a histogram analysis. An exemplary area of the cross section with its threshold analysis is shown in Figure 65.

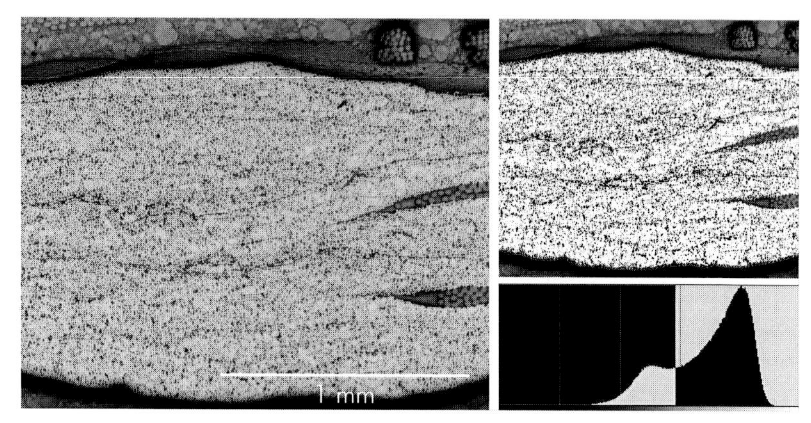

Figure 65. Microscopic image of Hybrid SMC (left); Image with threshold analysis resulting on the threshold controller (right)

X-ray microscopy (XRM) was used to validate the microscopic results. X-ray technology allows the non-destructive analysis of the outer and inner structure of three-dimensional test specimens. The ZEISS Xradia 520 Versa is equipped with a 160 kV microfocus tube. The X-ray source allows a maximum power of 10 W on the target. The optical setup consists of a scintillator and a 16-bit CCD 2024 x 2024 pixel camera. The two-stage geometric and optical magnification is complemented by a microscope objective system. The resulting low beam divergence enables a high resolution (< 1 μm) over a large measuring range (50 mm). Xradia also enables "time-of-flight based" phase contrast. The high resolution paired with the phase contrast enables the identification of unfilled resin areas between individual carbon fibre filaments. Focke et al. have described the Xradia 520 Versa in their studies with the help of fibre-reinforced plastic parts [146]. Due to the required resolution and the necessary phase contrast, the implementation of a XRM scan is time-consuming and cost-intensive. Therefore, X-ray microscopy was only used as a validation instrument to confirm the results obtained by light microscopy of the composite cross sections.
A section of a Hybrid SMC composite (10.0 x 5.0 x 2.0 mm) was taken for the validation scan with the XRM. It was manufactured with two textile reinforcements made by TFP with an unidirectionally arranged fibre architecture. Both TFP textiles are characterized by a bundle distance of 2 mm to each other and two carbon fibre layers on each other. The stitch density was set to 0.12 stitches per square millimetre. Both TFP textiles were positioned in the core of the Hybrid SMC composite The compression moulding was done with an isothermal temperature of 135 °C and a processing compression of 120 bar. The scan was

performed with a voltage of 60 kV. The distances between the X-ray source, object, and detector were adjusted to result in a voxel size of 2.6516 µm. The carbon fibre with a diameter of 7.0 µm was then sampled several times through the voxel size. The Nyquist-Shannon sampling theorem is fulfilled. The necessary settings (low voltage, large distances, sufficient exposure time, and number of single images (3200)) resulted in a scanning time of 9.5 hours. A section was taken from the same Hybrid SMC composite material and examined by light microscopy using threshold analysis.
Figure 66 shows a two-dimensional and cross-sectional image of a XRM scan as well as an image taken by the light microscope of the Hybrid SMC composite. At first glance, the images are similar to each other. The clear separation of the impregnated TFP fibre reinforcement with its constituents and the SMC cover layers can be clearly identified in both images. Nevertheless, and in contrast to the XRM scan, the two-dimensional image is just a cut-out of the complete composite structure and only offers information about the impregnation and FVC at the specific position of the Hybrid SMC composite.

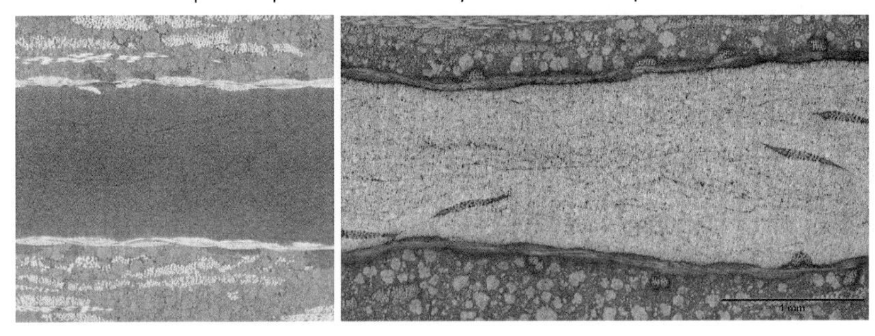

Figure 66. Hybrid SMC composite: Image of the XRM over-view scan (left) and light microscope image (right)

The data evaluation of the XRM measurement was performed with Volume Graphics 3.2. The void analysis is based on the threshold. Beside the position, the size of the void can be determined. In Figure 67, the void sizes are shown colour-coded.
The results of the XRM scans as well as the comparison with the light microscope images are presented in the following. First, the total defect content in the specimen is 2.5 %. This result is in correspondence with the void analysis of the threshold analysis considering the images by the light microscopy (1.8 %). Second, the centre of the fibre reinforcement shows more defects than the areas at the SMC cover layers. This was expected because the outer part of the fibre reinforcement is earlier impregnated than the core of the reinforce-

ment. Third, some defects are located at the sewing thread due to the inconsistency of the fibre reinforcement. Fourth, the majority of the voids are located along the fibre direction. Figure 67 clearly clarifies this observation. The image shows the defect analysis along the fibre direction. It can be assumed that the defects are located along the carbon fibre filaments. Due to this observation, it would be proven that the image analysis of cross-sections is a valid instrument to determine the void content within Hybrid SMC composites because each cross-section perpendicular to the fibre direction would lead to the similar void content because of the void arrangement along the fibres.

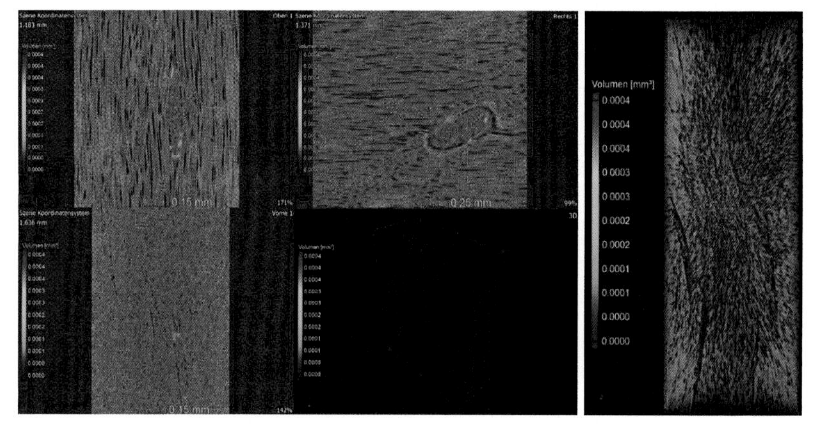

Figure 67. Void analysis inside the Hybrid SMC using XRM (high resolution – area of interest)

Using volume graphics, the arrangement of the defects can be described by the sphericity of the voids. The sphericity describes the surface-volume-ratio of a solid. The large volume voids (green – red; Figure 67) were not considered in the data analysis. These were mainly caused by the sewing threads. Figure 68 shows the volume of voids with its sphercitiy for the Hybrid SMC composites. A sphericity of 1 is equal to a spherical shape, lower sphericities are characterized by an elongated shape. The diagram demonstrates that greater void volumes have an elongated volume shape. Most of the time, these voids are located at the carbon fibre filaments. If occurring defects have a higher spherical shape, it is unfeasible to detect the void by the microscopic examination of cross-sections with image analysis. However, these kind of voids do not hinder the use of image analysis because these voids are small in contrast to the voids with an elongated shape.

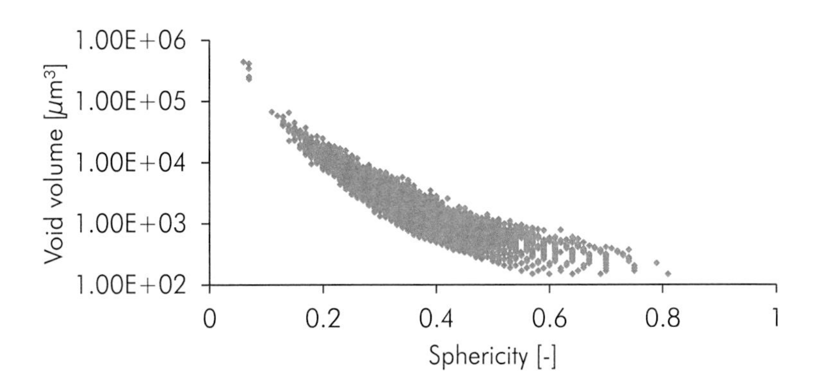

Figure 68. Void's sphericity in Hybrid SMC composites

5.1.3 Results

The major investigations are implemented with the double-sided fibre impregnation tests which represents a stacking sequence with two SMC cover layers and the TFP reinforcement in the core. In total, 135 Hybrid SMC laminates were manufactured and investigated. Therefore, more than 400 microscopic cross sections (including the investigations on curved fibre architectures (Chapter 5.2)) were investigated. Within the 135 Hybrid SMC laminates, the experiments were partially performed with a different hot press, a different TFP unit, or with a different mould. To make the individual test specimens comparable, just the specimens, which were manufactured with the same test equipment and manufacturing conditions, were compared to each other. In the following, the results of 39 Hybrid SMC composites were presented. The determined fibre volume content and the determined void content are shown in Figure 69.
It catches the eye that the void content acts contrary to the fibre volume content. The highest void content was found at 87.3 %. It is much higher in comparison to the second highest void content (31.0 %) that is why an outlier test is performed. The outlier test checks if each void content is a result of the individual parameters. If not, the void content can be a result of a mistake during the test implementation. The Grubbs's test for outliers is a method to determine outliers. One requirement to perform the Grubbs's test is the existence of a Gaussian distribution. Not only the Grubbs's test is based on a Gaussian distribution but most of the statistical tests are based on it. By ignoring this rule, false interpretations can occur.

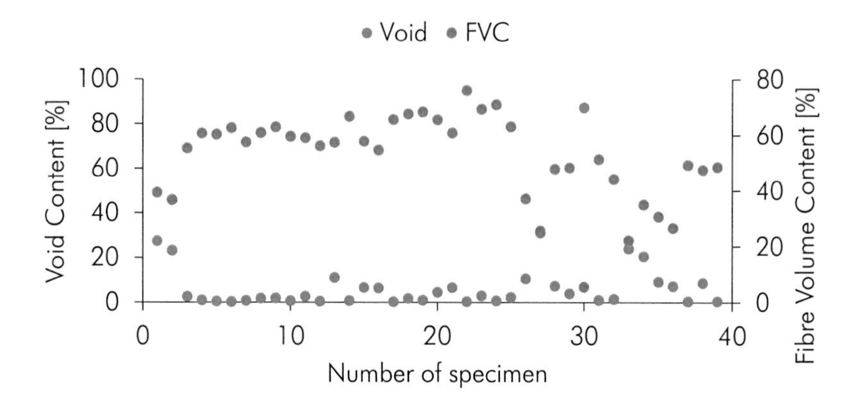

Figure 69. Void content and fibre volume content in Hybrid SMC composites

However, many statistical tests have a high robustness (e.g. t-test and F-test) that is why these tests may work anyway [147]. Nevertheless, the Gaussian distribution test should be always tested. The Gaussian distribution can be tested by a Q-Q-plot (Figure 70). A Gaussian distribution exists, when the standardized residuals of each measurement are close to the linear trend line. The theoretical quantiles assume the Gaussian distribution. Both the negative as well as the positive theoretical quantiles represent each 50 % of the Gaussian distribution. The standardized residuals represent the void content with respect to the average and standard deviation. The standardized residuals do not follow the linear trend line. Therefore, the following statistical evaluation requires a modification to consider the non-Gaussian distribution.

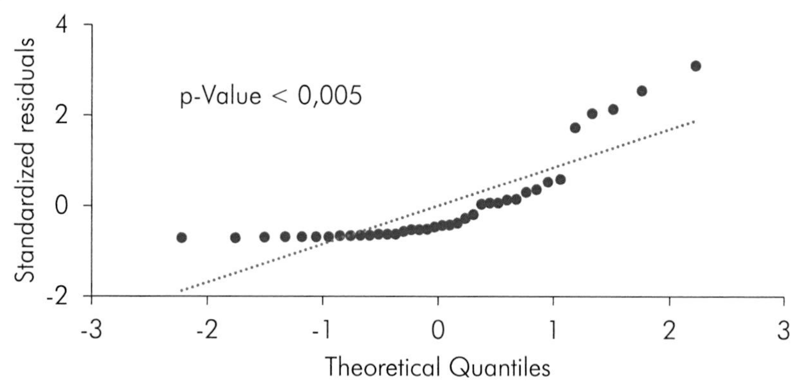

Figure 70. Q-Q-plot: No Gaussian distribution exists for the void content

This modification can be implemented by a Box-Cox-Transformation. The Box-Cox-Transformation transfers each result with an exponent λ to gain the best approximation of a Gaussian distribution. It works with the following equations:

$$f(\lambda) = \begin{cases} \frac{f^{\lambda}-1}{\lambda}, \text{ if } \lambda \neq 0 \\ \log y, \text{ if } \lambda = 0 \end{cases} \quad . \tag{5.3}$$

The ideal exponent was found in $\lambda = -0.13$ with a confidence interval between -0.28 and 0.01. Now, Grubbs's test for outliers can be performed with the transformed results of the void content. It was found that specimen no. 30, which represent the specimen with a void content of 87.3 %, is the only outlier. It is removed from the totality of results. Now, the average void content is 6.2 % and the standard deviation is 8.2 %. The average fibre volume content is 53.2 % but it is strongly influenced by the void content. By considering all specimens with a lower void content than 2.5 %, the fibre volume content is 58.0 %. In the next step, the main effect plots of individual parameters (Temperature (Figure 71), processing compression (Figure 72), stitch density (Figure 73), and reinforcement's thickness (Figure 74)) is presented. The main effect plots show the void content with regard to each individual parameter. A first insight is qualitatively given if a parameter has a special impact on the formation of voids during compression moulding. The dashed lines represent the average (6.2 %) value of the individual parameter.

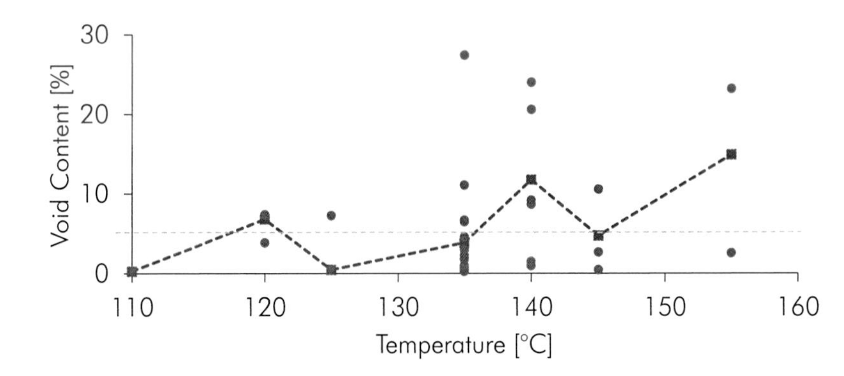

Figure 71. Main effect plot of the temperature

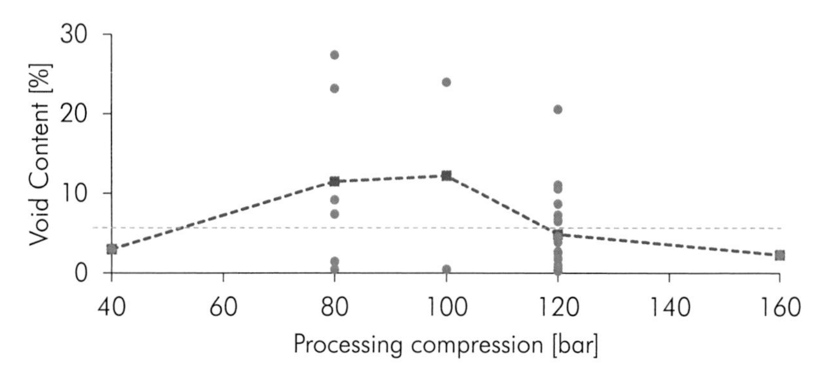

Figure 72. Main effect plot of the processing compression

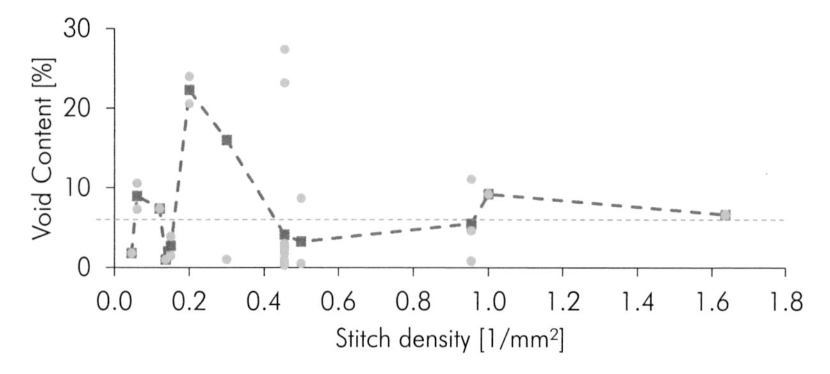

Figure 73. Main effect plot of the stitch density

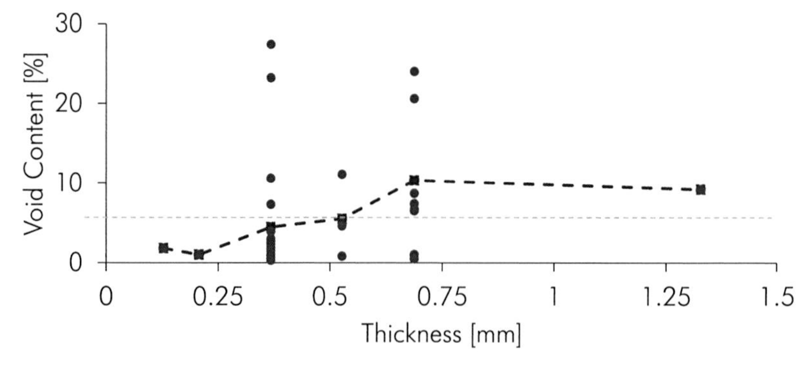

Figure 74. Main effect plot of the thickness

The main effect plot gives general information about the trend on the void content by changing one parameter but it does not consider potential coincidental effects. A slight trend can be found for the temperature and the reinforcement's thickness. An increase of the temperature and the reinforcement's thickness results in higher void contents. The processing compression shows the highest void contents at medium processing compressions. No real trend can be found for the stitch density. However, these observations are just a qualitative description of the impact of the individual parameters. With the help of statistical instruments, a deeper analysis can be implemented by a regression analysis with an additional significance test. The regression analysis checks if a correlation between the variables and the response exists. There are several methods to perform the regression analysis, for example a linear or multiple regression. A multiple regression analysis is chosen because of the four varied parameters at the same time. The void ratio is defined by κ. Note that the determined exponent λ of the Box-Cox-Transformation was considered. The following regression equation was found:

$$-(100\,\kappa)^{\lambda} = -2.036 + 0.355\,z_{half} + 0.0023\,\vartheta + 0.00819\,T - 0.00128\,P. \quad (5.4)$$

The coefficient of determination for the regression equation is low ($R_p^2 = 0.3911$). The coefficient of determination is equal to the squared correlation coefficient of the declared values and the variables to be declared [148]. By using a multiple regression analysis, it is the maximum correlation of the response with all linear parameter combinations [148]. Therefore, it describes how the response can be declared with the investigated variables. The coefficient of determination describes that only 39.11 % of the result are responsible for the resulting void content. This can be a result because of several reasons. First, linear regression was used which only is an approximated model. Effects of interaction were not considered. Second, further parameters can be involved which were not considered in the experiments. Nevertheless, the regression equation can be used to investigate the effect of the individual parameters.
The regression analysis goes on with the analysis of variance (ANOVA). With the help of the ANOVA, the significance of the parameters within the regression analysis can be tested. At first, the sum of squares (SQ), the mean squares (MQ), and the degree of freedom (DoF) which is the ratio of the sum of squares and the degree of freedom, are required to implement the ANOVA:

Sum of Squares (SQ): $$SQ = \sum_{i=1}^{c} (y_i - \bar{y})^2 \ , \quad (5.5)$$

Mean Squares (MQ): $$MQ = \frac{SQ}{DoF} \ . \quad (5.6)$$

The ANOVA was performed with a confidence coefficient of 95 % (significance level α_{sl} of 5 %). In the next step, an F-test is implemented which requires the definition of the null hypothesis H_0 and the alternative hypothesis H_1:

H_0: The individual parameters do not explain the result.
H_1: The individual parameters have an impact on the result.

The F-test is used to compare the variances of the samples to check if the model is valid for the population [147]. It is based on the F-distribution. The F-value is calculated considering the coefficient of determination, the sum of squares, the number of experiments, and the number of parameters. The F-value is compared with the reference F-value considering the significance level which is summarized in statistical literature [147]. If the calculated F-value is greater in comparison to the reference F-value, the null hypothesis H_0 can be declined and the alternative hypothesis H_1 can be accepted [147]. An alternative method for the determination of the significant effects is the comparison of the p-value with the significance level α_{sl}. The p-value can be described as the integral between the F-value and the end of the F-distribution. If the p-value is smaller than the significance level α_{sl}, the null hypothesis can be declined. Table 12 summarized the results of the ANOVA.

Table 12. ANOVA for the statistical evaluation of the void content

	DoF	SQ	MQ	F	p
Regression	4	0.366569	0.091642	5.30	0.002
Thickness	1	0.130745	0.130745	7.56	0.010
Stitch density	1	0.000025	0.000025	0.00	0.970
Temperature	1	0.235827	0.235827	13.64	0.001
Processing compression	1	0.022901	0.022901	1.32	0.258
Error	33	0.570610	0.017291		
Model	20	0.468495	0.023425	2.98	0.024
Error	13	0.102115	0.007855		
Total	37	0.937179			

Table 12 shows that the lowest p-value can be found for the temperature. Therefore, the temperature has a significant effect and the highest impact on the void content. An increase of the temperature will affect the fibre impregnation more than a change of any other parameter. Another significant parameter was found for the reinforcement's thickness. No significance was found for the processing compression and the stitch density. The processing compression is the driving force for the fibre impregnation that is why it is not negligible for the

Hybrid SMC compression moulding. The stitch density has the highest p-value and no effect on the fibre impregnation. Figure 75 shows the related Pareto chart which is in correspondence with the order of the individual p-values. The Pareto chart arrange the individual parameters in a sequence by considering its importance on the void content. It compares the parameters in a relative way to each other. The border effect is characterized by the red dashed line. It separates the significant and the non-significant effects. Significance is a mathematical-statistical quantity that must be subjected to a plausibility check in the specific case [147].

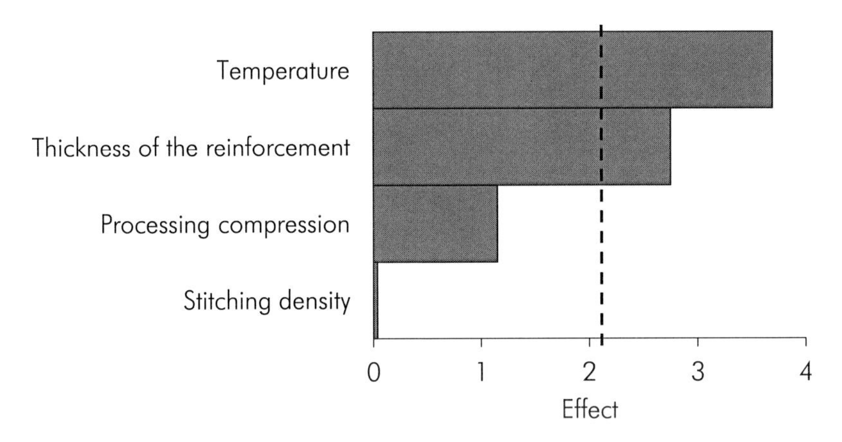

Figure 75. Pareto chart of the effects

The impact of the processing temperature was demonstrated for one-sided fibre impregnation tests. These tests were characterized with a textile reinforcement and only one SMC cover layer which is responsible for the total fibre impregnation. Only the temperature was varied and the other parameters were kept (processing compression, reinforcement's thickness, and stitch density) constant. The processing compression was 120 bar, the total thickness of the reinforcement was 0.735 mm, and the stitch density was 0.45 stitches per square millimetre. The curing cycle was individually chosen in dependence on the specific gelation point. Overall, ten specimens were tested and the determination of the void content was implemented by image analysis of the polished cross sections. The results are shown in Figure 76.

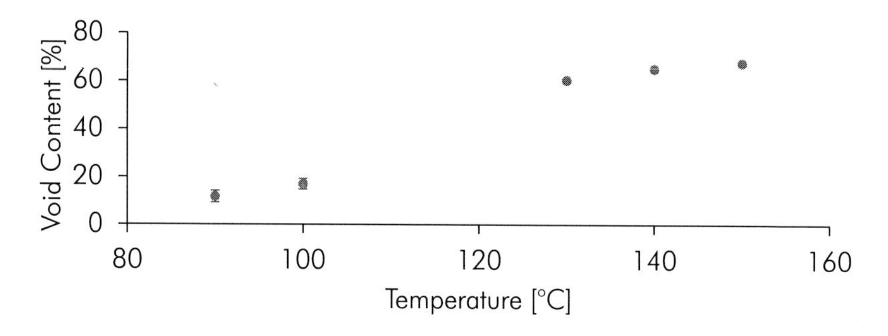

Figure 76. One-sided fibre impregnation of TFP textiles

The results prove the high impact of the temperature on the fibre impregnation. Lower viscosities can be used for a longer time and the gelation point is defered. Both effects are advantageous for the fibre impregnation. However, it is shown that even for a temperature of 90 °C, a total fibre impregnation cannot be realized by a one-sided fibre impregnation. In addition, the total cure of the Hybrid SMC composite takes a long time. If low temperatures are implemented within the compression moulding of Hybrid SMC composites, an appropriate method has to be developed to ensure the economic process conditions of the compression moulding process.

5.1.4 Alternative Approach: Non-Isothermal Curing Cycle

The experimental studies were conducted with an isothermal temperature and an isobaric processing compression. By using conventional process conditions for Hybrid SMC composites, the fibre impregnation can be completely realized with an adaption of the reinforcing textile regarding the thickness and permeability. If the gelation time is reached and the reinforcing textile is not finally impregnated, other solutions are required.
When the characteristics of the textile cannot be adapted, the process parameters need to be adapted. Because the processing compression does not have a significant impact on the formation of voids (Chapter 5.1.3), the process temperature can be lowered to extend the gelation time and to use a lower temperature- and time-dependent viscosity. In contrast to the processing compression, the temperature has a significant impact on the fibre impregnation. Figure 77 shows an exemplary temperature control for a compression moulding process to manufacture Hybrid SMC composites. Here, an impregnation temperature of 110 °C is chosen. At this temperature, the gelation time is only reached after five minutes which is an appropriate time to impregnate the single

carbon fibre filaments. The exact time can be predicted with the analytical impregnation model. At a temperature of 110 °C, the material requires more than 25 minutes to cure completely. Therefore, when the gelation point is passed, the temperature can be raised to a curing temperature to accelerate the material's curing. In Figure 77, a curing temperature of 140 °C was chosen. The disadvantage of this method is an extension of the total cycle time of SMC materials.

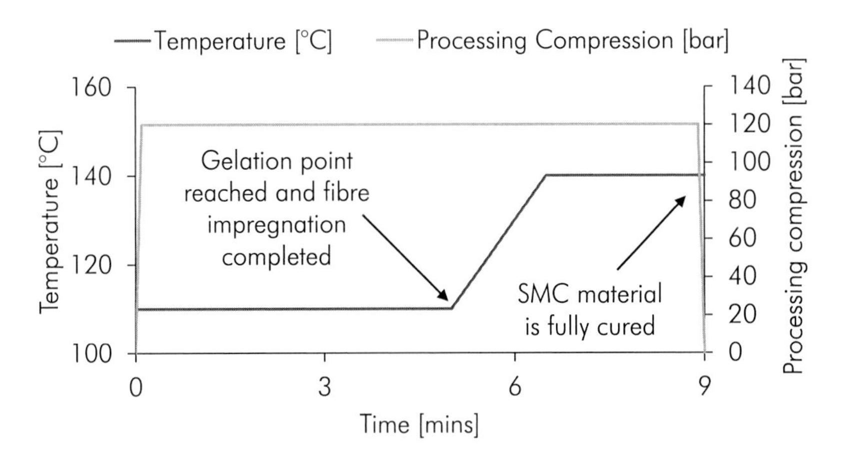

Figure 77. Non-isothermal curing cycle

The challenging task by using a non-isothermal curing cycle is the fast temperature increase to complete material curing as well as the temperature decrease to start the next compression moulding process. In addition, the glass transition temperature decreases by using a decreased curing temperature which can be important for applications. In accordance to the previous experiments (Figure 76), experiments were done by using the non-isothermal curing cycle in combination with the one-sided fibre impregnation. Eight Hybrid SMC composites were manufactured with TFP textiles containing carbon fibre reinforcements. The TFP textiles differ in the thickness to each other. The reinforcement's thickness was between 0.25 and 0.735 mm. A stitch density of 0.45 stitches per square millimetre was chosen. The results are shown in Figure 79. The void content increases with an increasing thickness of the fibre reinforcement because the impregnation length increases as well. In comparison to the results of Figure 76, the overall void content decreases with lower impregnation temperatures especially with regard to the relative low processing compression of 80 bar which was used for the experiments. In laboratory tests, the complete material cure was proven by DSC measurements. Figure 78 shows an exemplary polished cross section of a Hybrid SMC composite which was manufactured with the non-isothermal curing cycle.

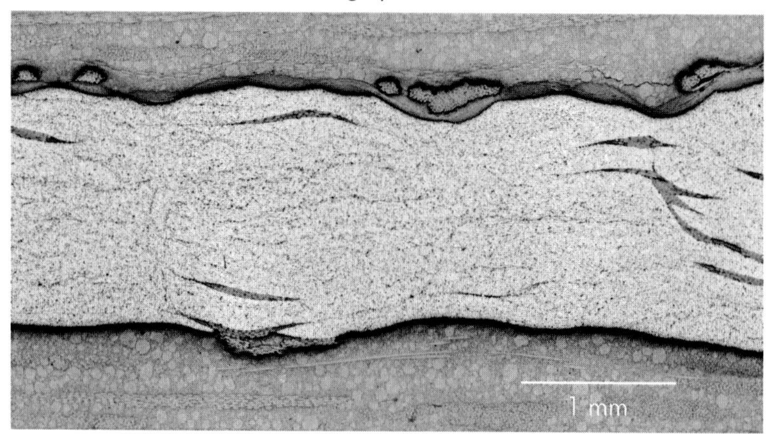

Figure 78. Exemplary microscopic image of a polished cross section (non-isothermal curing cycle)

Figure 77 has shown the temperature and processing compression which was used for the production of the Hybrid SMC composite. The thickness of the fibre reinforcement was 1.5 millimetre. Therefore, the fibre reinforcement was relatively thick in comparison to other fibre reinforcements within Hybrid SMC composites in this study. The use of the non-isothermal curing cycle has led to a void content of 2 %.

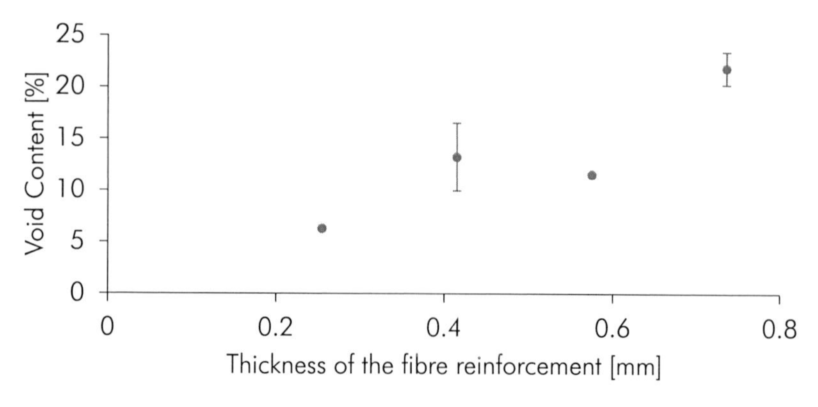

Figure 79. One-sided fibre impregnation of the TFP textiles by using a non-isothermal curing cyle

All in all, the non-isothermal curing cycle is an alternative to achieve good results for the fibre impregnation of thicker TFP textiles. The curing cycle extends in comparison to higher curing temperatures but the extension can be reduced by heating up. A further alternative could be the separation of the thick TFP reinforcement into two thinner TFP reinforcements.

5.2 Load Introducing Fibre Architecture

A load introducing fibre architecture describes the position of continuous fibre reinforcements, when mechanical loads punctually occur as in a bolted joint. The TFP technology allows a high flexibility in the fibre position. Therefore, a curved fibre design can be used to lay down the fibre around the punctual load introducing point to transfer the load within the structure. In contrast to conventional textile fabrics, a curved fibre design around a bolted joint avoids fibre damage by later drilling into the composite structure. In addition, punctual load introducing areas often have metallic inserts which provide lateral support for the curved fibre architecture and finally lead to an increase of the mechanical properties [22].
The following sections deal with load introducing fibre architectures of continuous fibre reinforcements. At first, some theories are described, which concern the fibre design, the fibre prestress, and the fibre impregnation by SMC materials. Then, experimental compression moulding tests are done and subsequently evaluated. In the end, qualitative conclusions sum up the chapter concerning the impregnation of load introducing fibre architectures.

5.2.1 Theoretics

Structural areas with punctual load introductions require an adapted fibre architecture to efficiently counteract mechanical loads. A loop structure is a promising method to design such areas. For this purpose, the fibre bundles have to fulfil several requirements to call up the full potential. A parallel loop structure is recommended to counteract only tensile stresses which was already demonstrated in the past [149], [150]. Other loads can be counteracted with a crossed loop design [22]. An alternative method is the implementation of two or more parallel loops to treat stresses. Furthermore, the position of the fibre bundles have to be close to the load introduction point because the highest stresses occur at the inner radius [22]. The position can be ensured with the application of fibre prestress. An experimental study illustrates the influence of the fibre prestress. Unidirectional carbon fibre bundles were processed by winding to a loop structure. After the manufacturing of the fibre reinforcement, the composite manufacturing was done with the resin transfer moulding technology.

Three different prestress levels were used within the winding process. The first carbon fibre loop was wound with almost no fibre prestress. The second and third carbon fibre loop were wound with an increasing fibre prestress:

$$F_1 < F_2 < F_3 \quad . \tag{5.7}$$

The impact of the fibre prestress was microscopically evaluated with polished cross sections at the vertex area (Figure 80). It can be seen that the height of the carbon fibre bundles at the vertex area decreases with higher fibre prestresses, so that the initial fibre volume content of 60.9 % (F_1) in this area increases to 68,5 % (F_2) and 81,8 % (F_3), respectively. This observation is in accordance with Lang et al. who made experimental investigations on loop structures [151]. Moreover, each carbon fibre bundle is clearly visible within the polished cross-section. The local fibre volume content does change within the cross sectional areas. The inner layers are characterized by higher local fibre volume contents in comparison to the outer carbon fibre layers (compare Figure 80, right side).

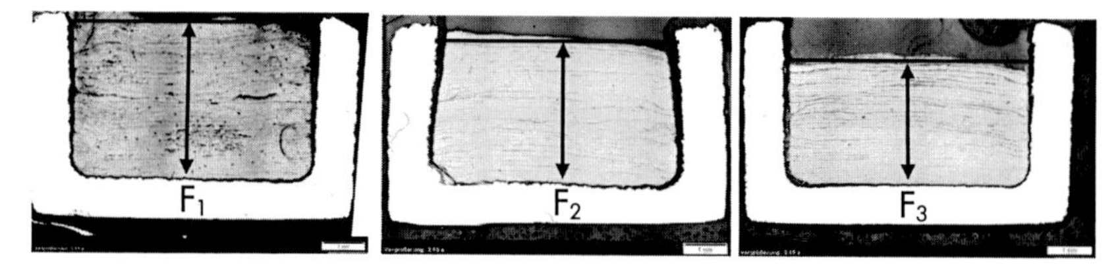

Figure 80. Impact of fibre prestress within the vertex area of a unidirectional carbon fibre composite loop

All in all, a good absorption for high mechanical loads can be realized when the carbon fibres lay at the inner radius. The implementation of fibre prestress is a good option to ensure the ideal position of the carbon fibre bundles. But fibre prestress leads to an increase of the fibre volume content that is why a reduction of the permeability would be the result. Lower permeabilities hinder a proper fibre impregnation and can be responsible for premature structural failure of carbon fibre loops. In addition, the fibre impregnation through the thickness cannot be realized because of the lateral support by the metallic elements. The fibre impregnation takes place from the outer to the inner side which leads to an in-plane impregnation. Curved fibre architectures within a TFP textile are characterized by an increasing amount of stitches by the sewing thread to keep the fibre position during the fibre curvature. The sewing thread can act as flow channels which enable the resin flow to the inner radius. All in all, there are other circumstances in comparison to elongated fibre architectures which have to be considered during the development of the design of load introducing areas in Hybrid SMC. An experimental test setup was designed to investigate the different parameters. In the end, qualitative statements are formulated concerning the impregnation of carbon fibre bundles in load introducing areas with metallic inserts.

5.2.2 Test implementation

To evaluate the impregnation of carbon fibre loops in load introducing areas, experimental tests have been conducted. These tests give information about the general impregnation behaviour of the load introducing fibre architectures by SMC materials. All in all, three adaptions are integrated to make qualitative statements regarding the design of fibre reinforced loop structures while compression moulding. These adaptions are:

1. Variation of fibre prestress
2. Increase of fibre bundle windings and use of greater metallic inserts
3. Compression moulding without metallic inserts

Like the unidirectional carbon fibre reinforcements, the loop structures are manufactured with the TFP technology. Regarding the manufacturing of the TFP textile, the same materials are used. A glass fibre woven fabric is used as a base material and the carbon fibre bundle is fixed with a polyester sewing thread (Figure 81).

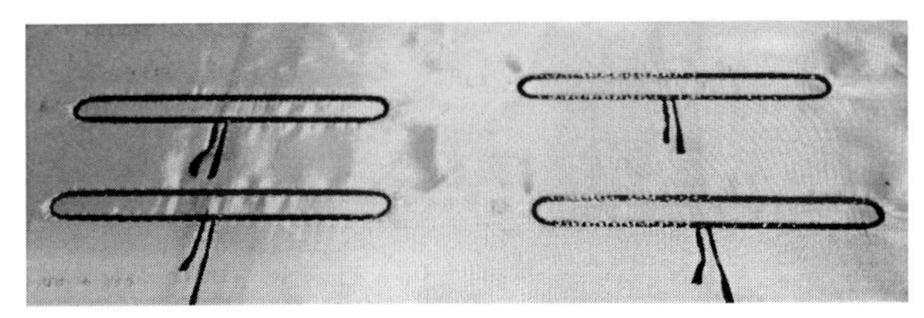

Figure 81. Carbon fibre bundles laid down in the shape of loops with the TFP technology

The carbon fibre loops are cut at the outer and inner contour after the TFP processing. Metallic inserts are put at the end of the carbon fibre loops. The mandrels are used within the mould to fix the metallic inserts. The carbon fibre loops are placed between two SMC layer before the compression moulding is started. In the following, the specific adaptions of the carbon fibre loops are described:

1. <u>Variation of fibre prestress</u>

The application of fibre prestress leads to an alignment of the carbon fibre reinforcement. The increase of the fibre volume content within the curved fibre architectures at the vertex area is assumed. The implementation can be done with active prestressing units inside the mould like it has been already presented for RTM moulds in the past [149]. The compression mould does not possess an active prestressing unit. However, the mould can be equipped with mandrels. Then, the fibre prestress can be implemented by the geometrical underdesigning of the carbon fibre loop structure. Both metallic inserts have a distance of 180.0 mm to each other. In the following experiments the carbon fibre loop was designed with a distance between both diameter centres of 177.0, 178.0, and 181.5 mm. The fibre prestress is a result of the fibre stretching due to the underdesigning of the carbon fibre loop. Each textile consists of six single carbon fibre bundle loops. The compression moulding was implemented with a temperature of 145 °C and a processing compression of 120 bar with a cycle time of five minutes.

2. Increase of fibre bundle windings and use of greater metallic inserts

In contrast to the first adaption, the amount of carbon fibre bundles was increased. The specimens of the first experiments just consist of six single carbon fibre loops. Now, instead of six, 20 carbon fibre loops are implemented to investigated the influence of thicker carbon fibre reinforcements. Therefore, greater metallic inserts were used. The inserts are characterized by an insert diameter of 28.0 mm instead of 22.0 mm. A temperature of 135 °C and a processing compression of 120 bar was chosen. No additional fibre prestress was used.

3. Compression moulding without metallic inserts

The experimental tests show the compression moulding of carbon fibre loop structures without any metallic inserts. A fibre impregnation through the thickness is realized. Nevertheless, a strong decrease of the mechanical performance is expected without any lateral support by the metallic elements [22]. The Hybrid SMC specimens contain a carbon fibre loop structure with 20 windings. The process was characterized by a temperature of 135 °C and a processing compression of 120 bar. Fibre prestress was used for the carbon fibre loops without any metallic inserts.

5.2.3 Qualitative Conclusion

In contrast to the evaluation of the unidirectional fibre architectures, the curved fibre architectures are evaluated in a qualitative way. Microscopic polished cross sections were prepared to describe the flow behaviour into the metallic insert and into the carbon fibre reinforcement. However, the individual void contents were not determined. The evaluation was restricted on a comparison between different Hybrid SMC specimens to identify differences within the fibre impregnation. Only one single material parameter was changed in the compression moulding process to evaluate the impact of this parameter.

1. Variation of fibre prestress

The fibre prestress is set with the geometrical underdesigning of the carbon fibre loop between the fixed mandrels. A microscopic polished cross section of the vertex area of an underdimensioned carbon fibre loop is shown in Figure 82.

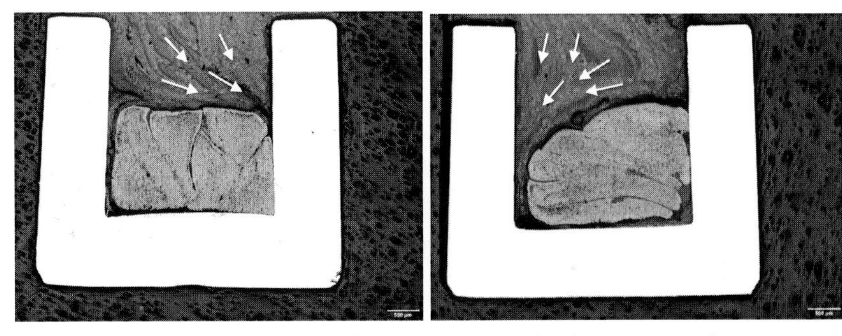

Figure 82. Vertex areas carbon fibre loop in a Hybrid SMC way of construction (without (left) and with prestress (right))

At the beginning of the compression moulding process, the material flows into the metallic insert. The material takes the way with the lowest resistance. In each case, this area is the transition area between the TFP textile and the metallic insert. This is noticeable due to the alignment of the materials within the SMC due to compression moulding. The SMC material flows along the metallic insert. Then, the fibre impregnation of the carbon fibre reinforcement partially takes place from the back. The material uses the sewing thread as a flow channel which can be noticed by the flow channels between the carbon fibre bundles. Overall, the fibre impregnation is slightly better for the specimen with additional fibre prestress in contrast to the specimen without any prestress. However, the carbon fibre loop without any prestress shows greater uniformity than the carbon fibre loop with additional prestress. This was noticed for several specimens. The fibre volume content is similar in both specimens. No qualitative difference can be noticed. The flow channels between the carbon fibre bundles and along the sewing thread are slightly smaller at the carbon fibre loop with prestress in comparison to the other specimen because of higher compression stresses which are caused by the fibre prestress.

2. Increase of the fibre bundle windings

The following experiments are characterized by an increase of fibre bundles within the loop structure from an amount of six to 20 carbon fibre loops. Thereby, the impregnation length increases because the reinforcement's thickness increases as well. The metallic insert was adapted due to a lengthening of the lateral support to ensure that the additional fibre bundles have enough space inside the metallic insert. Figure 83 shows three polished cross sections of one metallic insert at different positions of the metallic insert.

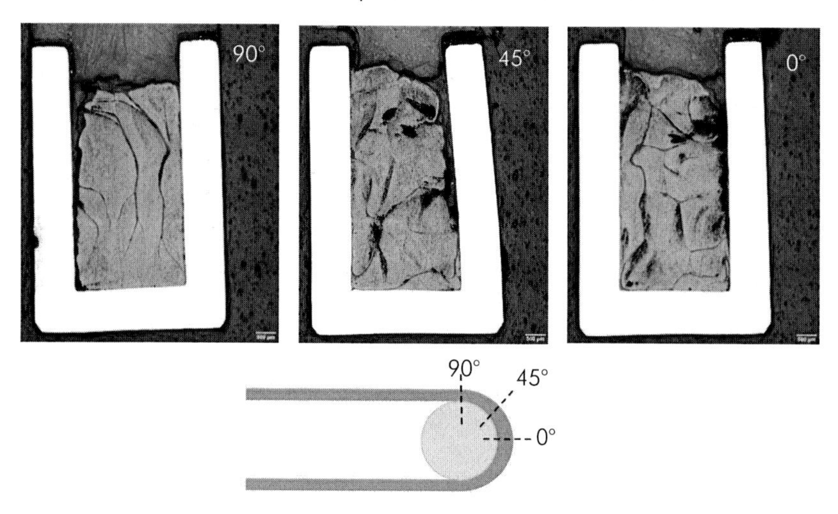

Figure 83. Development of the fibre impregnation at the transition area (start of metallic insert – 90° (left); curved cross section – 45° (middle); and vertex area (right))

The first cross section shows the transition of the unidirectional fibre architecture to the curved fibre architecture. The next polished cross section was taken at an angle of 45° at the metallic insert. The last one shows the polished cross section at the vertex area.

The best fibre impregnation results are observed in the transition area to the unidirectional fibre architecture and the metallic insert (90°). The other two images show an increasing trend regarding the void content. Again the flow channels of the SMC materials are clearly visible. This effect is caused by the existence of the metallic inserts because the processing compression does not directly act on the carbon fibre reinforcement but it partially acts from the SMC material which is pushed into the fibre reinforcement. Especially the cross section of the vertex area consists of voids at the outer area of the carbon fibre reinforcement. It seems that the entrapped air is unable to escape at the vertex.

Nevertheless, it was shown that inner carbon fibre bundles can be impregnated by the SMC resin even if the impregnation length is high but dry spots still remain. The reason is the existence of flow channels which are located between the TFP reinforcement and the metallic insert most of the time. A complete fibre impregnation can potentially be achieved by choosing the right material and process parameters. In this study, the impact of additional fibre prestress on the fibre impregnation of the thicker carbon fibre bundle with its larger metallic element was not tested.

3. Compression moulding without metallic inserts

Compression moulding without any metallic inserts at load introducing areas leads to an impregnation of the fibre reinforcement through the thickness instead of the sideward impregnation. However and in general, the construction of FRP without considering metallic inserts should be avoided, especially when high stresses occur because the fibre loop design has no lateral support and would lead to premature inter-fibre failure [22]. Figure 84 shows an exemplary image of the fibre impregnation of the vertex area of a carbon fibre loop without any metallic inserts. The inner part of the carbon fibre loop has a poor fibre impregnation. A part of this area is characterized by completely non-impregnated carbon fibre bundles. Non-impregnated areas at the inner radius would lead to premature failure. The poor fibre impregnation is a result of fibre prestress and the resulting higher local FVC which mainly acts on the inner carbon fibre bundles. A reduction of the prestress can lead to a better fibre impregnation. In addition, the carbon fibre bundles were flattened due to the high processing compressions. Overall, the fibre impregnation through the thickness has no positive effect. The outer fibre bundles have a better impregnation but a complete fibre impregnation cannot be realized. The single filaments in the middle have non-impregnated areas in the middle of the carbon fibre bundle that is why the impregnation is poor on micro scale. The flow paths between the individual carbon fibre bundles are clearly visible.

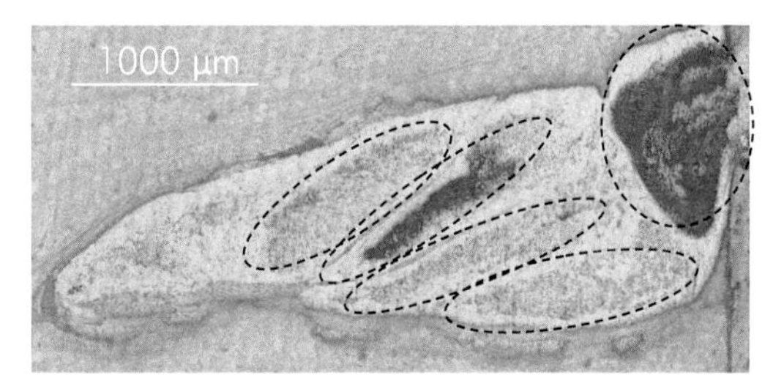

Figure 84. Fibre impregnation of curved fibre architecture without any metallic load introduction elements

All in all, the presented qualitative investigations just give an insight about the complexity of the fibre impregnation at load introducing areas. The investigations have shown that quality assurance is an important topic to guarantee the right position of the fibre reinforcement. Therefore, non-expected flow channels can be avoided. But even if the processing compression is not able to act through the thickness because of the lateral support of the metallic insert, a convenient fibre impregnation is enabled (Figure 82). Nevertheless, there are issues regarding the fibre impregnation which have to be considered during the development of processing Hybrid SMC composites with curved fibre architectures (Figure 83, Figure 84). These issues concern the textile and the insert design, the implementation of the process, and the choice of the right fibre prestress to call up the maximum mechanical performance at load introducing areas.

5.3 Alternative way of Production: Pre-Impregnation of TFP textiles

An alternative way of production is the use of an additional process step to impregnate the fibre bundles within TFP textiles. This has been already demonstrated by several researchers in the past [80]. However, it leads to an additional manufacturing step which hinders the economic potential of the SMC compression moulding process. Figure 85 demonstrates an alternative way of production when the pre-impregnation of the carbon fibre textiles is preferred.

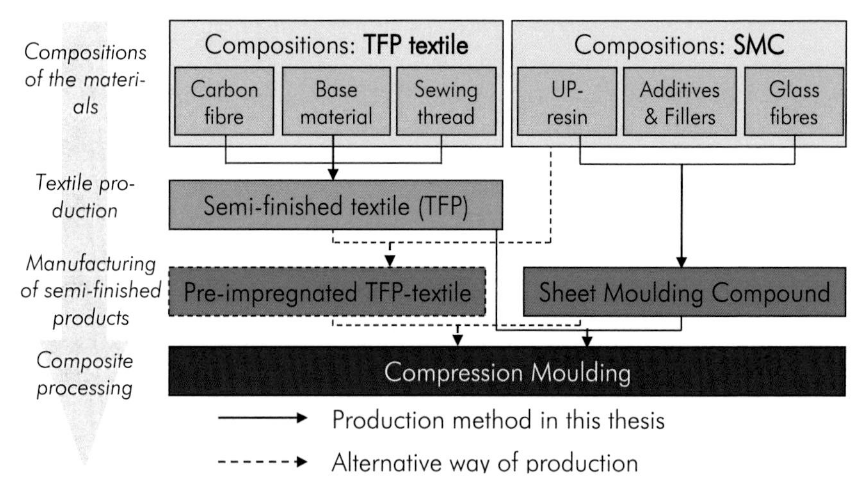

Figure 85. Manufacturing chains for processing Hybrid SMC

The pre-impregnation of a TFP textile can take place on a conventional SMC manufacturing lane (Figure 86). This production method requires an endless base material which merges the individual TFP textiles. Thereby, a continuous feed of the textile into the SMC impregnation lane was ensured. Modern TFP units are able to manufacture on endless base materials. The same resin mixture can be used to manufacture the semi-finished SMC as the resin mixture in the SMC material. But if necessary, the resin mixture can be adapted. Then, for example, the resin mixture can be made without any filler contents which may lead to an increase of the viscosity and hinder the fibre impregnation.

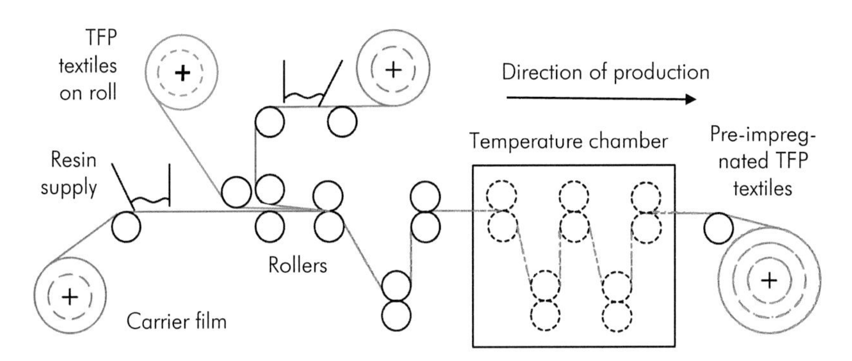

Figure 86. Adapted SMC impregnation lane for the fibre impregnation of TFP textiles

After the pre-impregnation of the TFP textiles was implemented, the fibre impregnation was investigated. For this purpose, several TFP textiles were manufactured on one continuous base material. The TFP textiles are characterized by eight layers of carbon fibres. This amount of carbon fibre layers has shown poor impregnation results, when a reinforcing textile without any pre-impregnation was used. Then, the continuous TFP textile was fed into a SMC impregnation lane with a tape speed of 2 m/min (Figure 87). The resin quantity was set to 240 g/m^2 and applied on the carrier films.

Beside the UP resin and the required percentage of styrene, an additional contrast medium was integrated to investigate the fibre impregnation. Thereby, a stronger contrast between the fibres and the resin shall be achieved. The contrast medium SACHTOPERSE HU-D supplied by Sachtleben Chemie GmbH was used. It is an untreated synthetic barium sulphate with a density of 4.5 g/cm^3 and a particle size smaller diameter than 1.0 micrometre. This particle size allows the evaluation of the fibre impregnation in all areas, even between the single carbon fibres. The ratio of the contrast medium to the overall resin mixture was 2.0 %. An additional calendar pressure can be applied to improve the impregnation. Within this impregnation of the TFP textiles, the calendar pressure was set to 0.15 MPa. Optionally, the calendar rolls can be heated to lower the resin's viscosity. However, this temperature control system was not used.

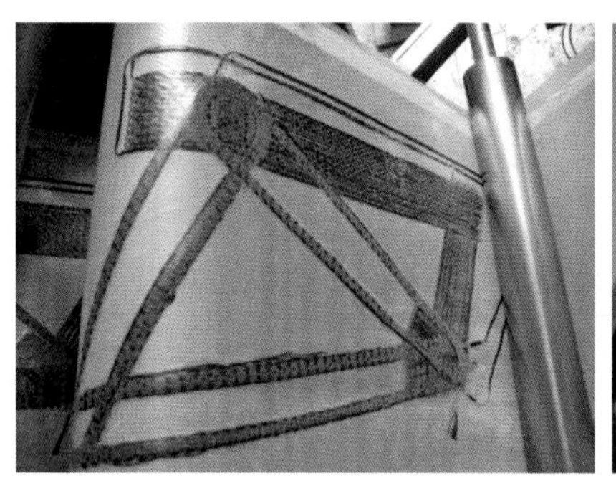
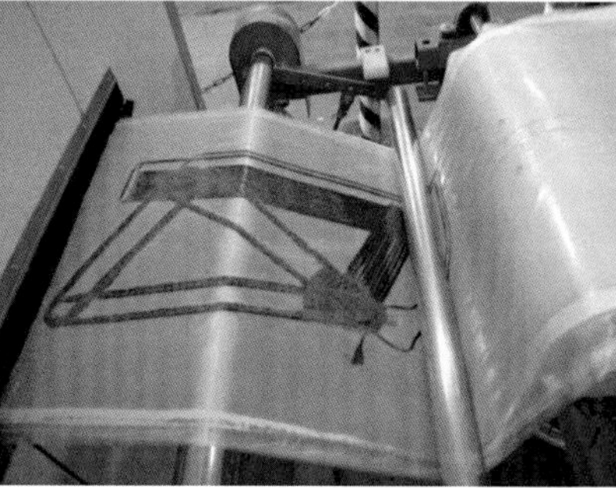

Figure 87. Pre-impregnation of a TFP textile with a SMC impregnation lane

The impregnated TFP textiles were cut into small pieces to prepare the specimens for the investigation with a XRM because the resolution is dependent on the size of the specimen. Therefore, a size of 1.0 x 1.0 x 1.0 cm was used which corresponds to four impregnated cut outs on each other considering the thickness. The scans were done with the 'Xradia 520 Versa' by Zeiss. After the pre-impregnation of the textile was realized, the single carbon fibre layers of each reinforcing TFP textile can be identified due to the flow front of the resin (Figure 88).

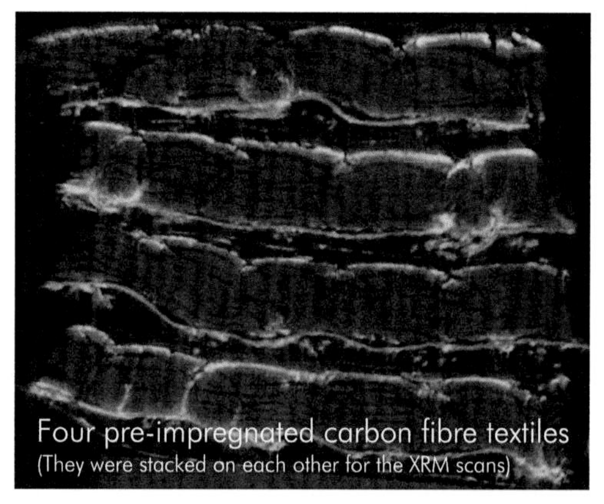

Figure 88. CT scan of four pre-impregnated carbon fibre textiles

The image clearly shows the contrast medium on the surface of the TFP textile. Just in individual cases, the contrast medium flow through the carbon fibre reinforcement. Partially, it can be seen in the direction through the thickness. Here, the sewing thread is located. The sewing thread acts again as a flow channel. The permeability is locally higher because the resin flows along the sewing thread (Figure 89). Despite the low particle diameter of the contrast medium, it is not able to go between the single carbon fibres. It seems, that the particles of the contrast medium merges to conglomerates which hinder the further distribution within the carbon fibre reinforcement.

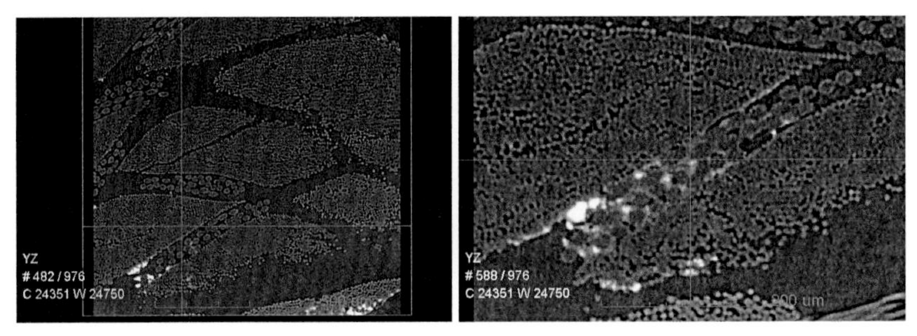

Figure 89. Contrast medium at the flow channels at the sewing thread

Despite the unequal distribution of the contrast medium within the UP resin, a low void content can be noticed. In contrast to the entrapped air, the resin has another grey scale value. Carbon fibres and the polyester sewing thread have a similar grey scale value but a differentiation is possible because of the fibre diameter and the position. Furthermore, the progress of the fibre impregnation can be seen. At first, the resin flows along the sewing thread and surrounds each carbon fibre bundle (macro impregnation). Due to the partial compaction of the calendar rolls, each carbon fibre bundle was impregnated from outside to inside (micro impregnation). The single carbon fibres of each fibre bundle were finally impregnated by the resin. The subsequent compression moulding process was not implemented because the full fibre impregnation of the TFP textile has already been proven.

All in all, the impregnation of the continuous carbon fibre bundles can be ensured with a pre-impregnation of the TFP textile. However, the pre-impregnation of TFP textiles have demonstrated disadvantages as well. At first, a further production step certainly leads to higher production costs. The economical evaluation is always an essential part regarding the decision, if a new technology will replace the conventional method. Furthermore, the impregnation of flexible, load-optimized fibre architectures within a TFP textile results in fibre deformations due to the calendar rolls within the impregnation lane (Figure 90).

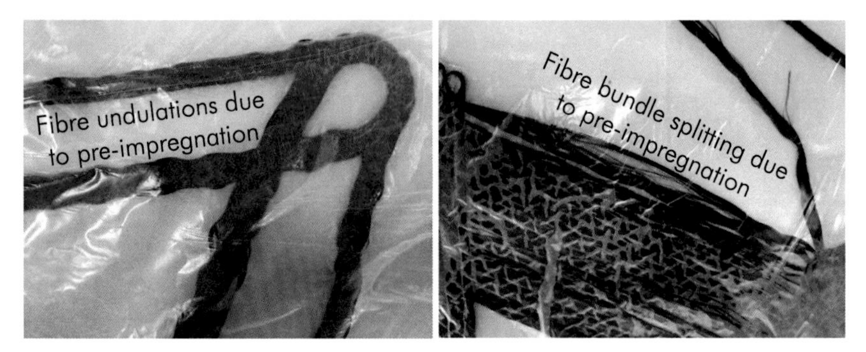

Figure 90. Fibre deformations within TFP textiles after the pre-impregnation process

The mechanical potential of the fibre reinforcements decreases because of these reasons. To counteract the fibre deformations of load-path optimized carbon fibre reinforcements, additional fibre prestress is necessary during compression moulding. The implementation of fibre prestress has to be considered in the construction of the mould. In the experiments it was already demonstrated that a complete fibre impregnation can be achieved after the compression moulding of non-impregnated carbon fibre reinforcements.

6 Thermo-Chemo-Rheological Impregnation Model

This chapter describes the development of an impregnation model for Hybrid SMC composites. In a first step, the development of the impregnation model is described. The dependencies of the individual parameters are shown which influence the fibre impregnation. The individual material and process conditions are implemented to the impregnation model. Afterwards, a validation of the impregnation model is fulfilled by considering the results of the experimental tests.

6.1 Model development

The approach for the development of the impregnation model is based on the fluid dynamics. There are two main approaches which describe the movement of a fluid's particle. The approaches are called Eulerian and Lagrangian description [152]. The Eulerian approach observes a specific area and determines, when the fluid is passing it. The Lagrangian approach follows a specific particle within the fluid at any place and any time. It tracks the particle as a function of time [152]. Both approaches use the material derivative which is defined in the following equation:

Material derivative:

$$\vec{a} = \frac{D\vec{V}}{Dt} = \frac{\partial \vec{V}}{\partial t} + (u\frac{\partial \vec{V}}{\partial x} + v\frac{\partial \vec{V}}{\partial y} + w\frac{\partial \vec{V}}{\partial z}) \quad . \tag{6.1}$$

The material derivative describes the acceleration by the derivation of the fluid's velocity with respect to the particle's movement in the coordinate system and the time. Darcy's Law describes the fluid's velocity:

$$V = \frac{K}{\eta}\frac{\Delta P}{\Delta z} \quad . \tag{6.2}$$

In the case of Hybrid SMC compression moulding, the fluid is the resin inside the SMC material which flows through the carbon fibre reinforcement. The carbon fibre reinforcement represents the porous medium. Due to the exclusive flow through the thickness of the reinforcement as well as a mould coverage of 95 %, the resin does only flow through the thickness. By definition, this is the z-direction. Therefore, equation 6.1 reduces to a one-dimensional case. In addition, the vectorial notation can be eliminated:

Material derivative (1D) $$a_z = \frac{DV}{Dt} = \frac{\partial V}{\partial t} + w\frac{\partial V}{\partial z} \quad . \tag{6.3}$$

The velocity of SMC materials decreases because of an increase of the material's viscosity with respect to time. Therefore, the fibre impregnation takes place with an unsteady flow ($\frac{\partial V}{\partial t} \neq 0$). Due to the stitching pattern, which can act as flow channels through the thickness, the flow through the fibre reinforcement is non-uniform ($\frac{\partial V}{\partial z} \neq 0$). A faster flow of the resin exists in the areas of the sewing thread. However, the impregnation model does not make a difference between the impregnation along the sewing thread and the impregnation perpendicular the carbon fibre reinforcement. Therefore, the impregnation model works with the assumption of a uniform resin flow through the thickness ($\frac{\partial V}{\partial z} = 0$). Figure 91 illustrates the difference between the uniform flow front within the impregnation model and the non-uniform resin flow in reality.

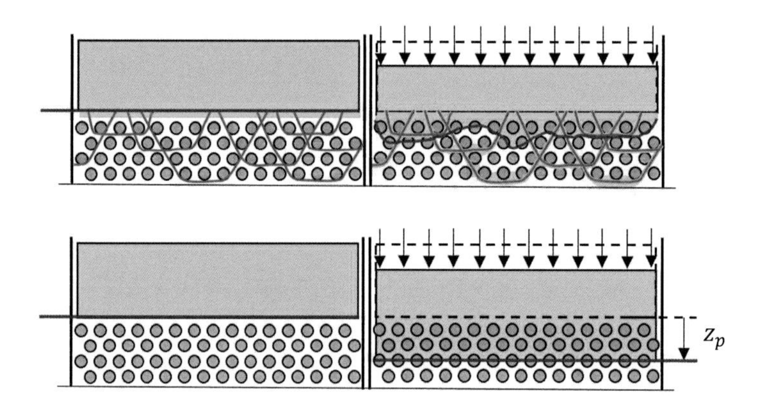

Figure 91. Flow front within Hybrid SMC compression moulding: Reality (top) vs. Model (bottom)

Moreover, this assumption includes that the impregnation model does not make a difference between the impregnation on macro, meso or micro scale. In contrast to the fibre impregnation on micro scale in reality, the model works with a mixed resin flow front on macro scale which already considers the micro impregnation with an averaged value. This essential difference is illustrated in Figure 92.

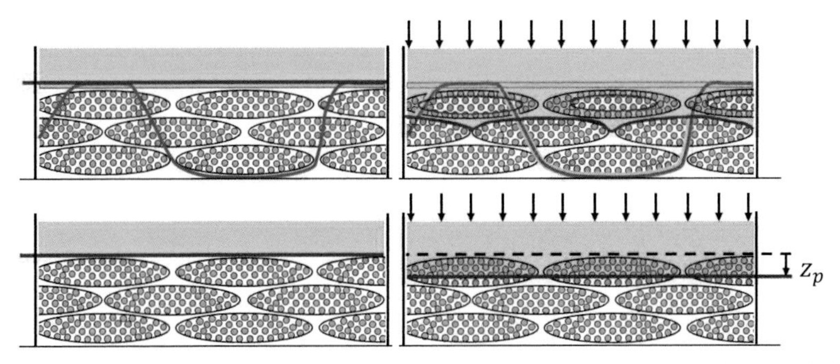

Figure 92. Impregnation in reality (top) and mixed impregnation in the model (bottom)

By using a uniform and an unsteady resin flow through the thickness, the equation for the material derivative can be simplified. The fluid's velocity through the thickness of the carbon fibre reinforcement just changes with respect to time:

Material derivative (Uniform and unsteady flow) $$a_z = \frac{DV}{Dt} = \frac{\partial V}{\partial t} \quad . \tag{6.4}$$

A further simplification can be used because of the symmetric stacking sequence. The non-impregnated continuous carbon fibre textile is covered by the outer SMC material. Therefore, both SMC cover layers have to flow to the middle of the carbon fibre reinforcement. This is the half of the total thickness of the reinforcement (Figure 93). Note, that Figure 91 and Figure 92 has already considered the symmetrical case.

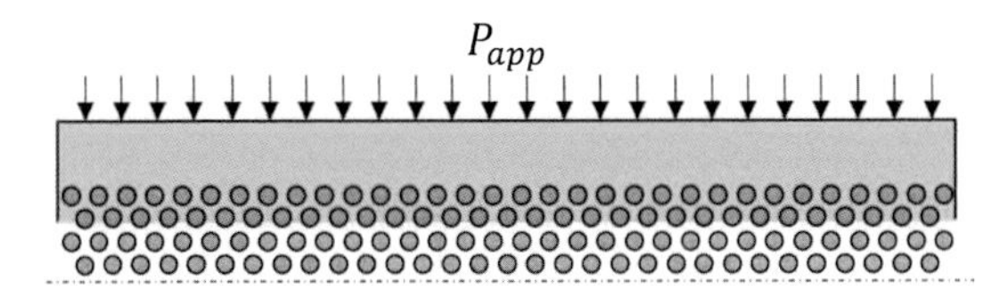

Figure 93. Consideration of the symmetrical stacking sequence in Hybrid SMC

After the simplification of the material's derivative as well as the consideration of the symmetrical conditions was described, the analytical equations are developed. The Lagrangian approach tracks a specific particle with respect to time. The particles are tracked at the flow front of the SMC material because it gives information about the flow length (Figure 91, red line). The particles at the flow front are defined as z_{ff} as a function of time:

$$z_{ff} = f(t) \quad . \tag{6.5}$$

The velocity at the flow front particles is the derivative of the particle's position:

$$v_p = \frac{df}{dt} = \frac{dz_{ff}}{dt} \quad . \tag{6.6}$$

The velocity is described by Darcy's Law and can be substituted into the equation:

$$v_p = \frac{dz_{ff}}{dt} = \frac{K_z}{\eta} \frac{P_{app}}{0.5\, z_{total}} <=> \frac{K_z}{\eta(t)} \frac{P_{app}}{0.5\, f} = \frac{df}{dt} \quad . \tag{6.7}$$

Now, the equation can be separated with respect to the derivatives:

$$\frac{K_z}{\eta(t)} P_{app} dt = 0.5 f df \quad . \tag{6.8}$$

At this point, a definition of the boundary conditions is required. The boundary conditions contain information about the starting and end conditions with respect to time and location. The beginning of the process is defined at the transition between the maximum compaction is achieved after starting the compression moulding process and the first carbon fibre filaments are impregnated [109]. The top of the compacted reinforcement is defined as the neutral axis. At this point, both the position and the time is zero. The end of the process is limited by the gelation point. After this time, no more fibre impregnation can take place. The end position of the flow front is limited at the middle of the fibre reinforcement because of the symmetrical stacking sequence. The boundary conditions are summarized in Table 13.

Table 13. Start and end conditions of the impregnation model

	Time	Position
Start of the process	$t = 0$	$z = 0$
End of the process	$t = t_{gel}$	$z_{ff} = ?$

Now, equation 6.8 is integrated and solved with respect to z_{ff} by using the boundary conditions from Table 13 to define the limits the integrals:

$$\int_0^{t_{gel}} \frac{K_z}{\eta(t)} P_{app} dt = \int_0^f 0.5 f df \quad . \tag{6.9}$$

$$K_z P_{app} \int_0^{t_{gel}} \frac{1}{\eta(t)} dt = 0.5 \int_0^f f df \quad . \tag{6.10}$$

$$K_z P_{app} \int_0^{t_{gel}} \frac{1}{10^{mt+n}} dt = 0.5 \int_0^f f df \quad . \tag{6.11}$$

$$K_z P_{app} \int_0^{t_{gel}} \frac{1}{10^{mt+n}} dt = 0.5\,(0.5f - 0) \quad . \tag{6.12}$$

$$K_z P_{app} \int_0^{t_{gel}} \frac{1}{10^{mt+n}} dt = 0.25 f^2 \quad . \tag{6.13}$$

$$f = z_{ff} = \sqrt{4\, K_z P_{app} \int_0^{t_{gel}} \frac{1}{10^{mt+n}} dt} \quad . \tag{6.14}$$

The permeability through the thickness K_z is described by an analytical calculation by Gebart using an quadratic fibre arrangement. By using Gebart's analytical solution, equation 6.14 transforms to the following equation:

$$f = z_{ff} = \sqrt{P_{app} \frac{64 r_{cf}^2}{9\pi\sqrt{2}} \left(\sqrt{\frac{\frac{\pi}{4}}{\varphi_f}} - 1 \right)^{\frac{5}{2}} \int_0^{t_{gel}} \frac{1}{10^{mt+n}} dt} \quad . \tag{6.15}$$

Now, the flow front z_{ff} can be calculated for each process condition within Hybrid SMC compression moulding with respect to the temperature, the processing compression, and the fibre diameter. The processing compression P_{app} is determined by the hot press. In this study, a typical carbon fibre diameter was used ($2\, r_{cf} = 7\, \mu m$). The process temperature determines the missing variables m, n and t_{gel} which were already determined by the prior viscosity measurements of the SMC material:

$$m = 6.8 \cdot 10^{-7} T^3 - 0.00019\, T^2 + 0.0179\, T + 0.5566 \quad , \tag{6.16}$$

$$n = -2.1 \cdot 10^{-6} T^3 + 0.00082\, T^2 - 0.1107\, T + 8.2165 \quad , \tag{6.17}$$

$$t_{gel} = 1 \cdot 10^6\, e^{-0{,}075\, T} \quad . \tag{6.18}$$

The polynomic function for the m-value is negative for temperatures between 96 and 107 °C which leads to strong impact to the solution of the integral. Mathematical methods can fix the problem but it would lead to an inconsistent course of the graph. Moreover, these temperatures are not preferred during compression moulding that is why this temperature range is neglected for the impregnation model. The use of the impregnation model is limited to temperatures between 110 and 160 °C.
The final degree of impregnation is the ratio between the flow front length z_{ff} and the half of the carbon fibre reinforcement's thickness z_{half} because of the symmetrical stacking sequence. The void ratio is described by the variable κ:

For $z_{ff} < \frac{1}{2} z_{total}$:
$$\kappa = 1 - \frac{z_{ff}}{z_{half}} \quad . \tag{6.19}$$

For $z_{ff} \geq \frac{1}{2} z_{total}$:
$$\kappa = 0 \quad . \tag{6.20}$$

The calculation of the total reinforcement's thickness was presented with equation 4.21. It is dependent on parameters during the manufacturing of the TFP textile. These parameters are the bundle distance and the amount of carbon fibre layers. In the following, equation 6.15 in combination with equations 6.19 and 6.20 are used for the validation of the impregnation model.

6.2 Validation and Fitting

This chapter is about a comparison between the developed impregnation model and the experimental results. It gives information about the accuracy of the impregnation model. If necessary, the impregnation model will be fitted to the experimental results. At first, a comparison of the experimental results is done with the impregnation model by using the average fibre volume content of 58 %. The results of the comparison can be shown in Figure 94. It catches the eye that there is a high deviation between both results. The void content is overestimated by the impregnation model. Most of the time, the void content of the impregnation model is higher than 60 % but the experimental results have shown that the void content is mostly lower than 10 %. Only three results of the impregnation model are close to the experimental results.

Figure 94. Comparison of the measured void content and the void content by the impregnation model

Because of the overestimation of the impregnation model, an empirical parameter was developed and added. Figure 95 shows the comparison between the experimental results and the impregnation model with an empirical parameter of $\xi = 4.87$. It catches the eye that the model represents a better fitting with this empirical parameter. By using the least square method, the overall deviations were reduced by more than 92 %.

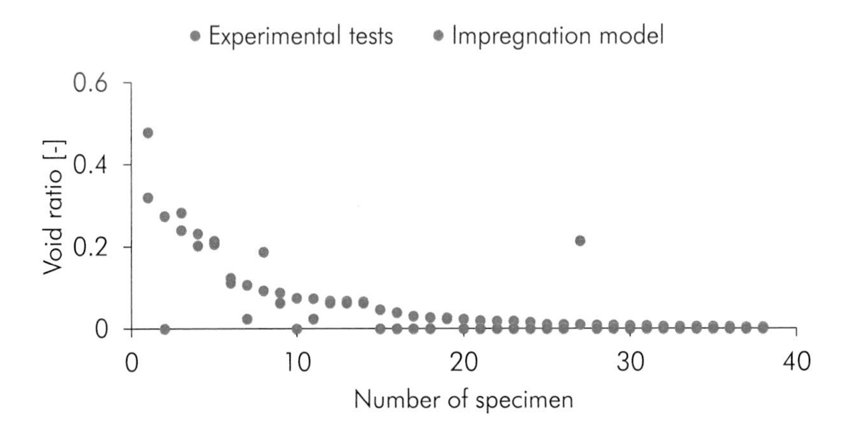

Figure 95. Comparison of the measured void content and the void content by the impregnation model with an empirical parameter

The second specimen shows a higher deviation between the experimental measurement and the impregnation model. It was manufactured with a temperature of 135 °C, a processing compression of 80 bar, a reinforcement's thickness of 0.367 mm and a stitch density of 0.06 stitches per square millimetres. Due to the moderate temperature as well as the moderate thickness of the reinforcement, the experimental determined void content surprises even if a processing compression of 80 bar was used. For example, specimen 17 was manufactured with the same parameters except the processing compression (40 bar) and has led to a void content of 3.0 %. Therefore, the second specimen is treated as an outlier which was not statistical proven. The empirical parameter is added to the permeability's analytical calculation because of the use of analytical permeability models which were not experimental confirmed and a lack of information regarding the permeability still exists:

$$K_z = 4.87 \frac{16}{9\pi\sqrt{2}} \left(\sqrt{\frac{\frac{\pi}{4}}{\varphi_f}} \right)^{\frac{5}{2}} r_f^2 \quad . \tag{6.21}$$

Finally, it is added to the impregnation model:

$$z_{ff} = \sqrt{P_{app} \frac{311.68\, r_{cf}^2}{9\pi\sqrt{2}} \left(\sqrt{\frac{\frac{\pi}{4}}{\varphi_f}} - 1 \right)^{\frac{5}{2}} \int_0^{t_{gel}} \frac{1}{10^{mt+n}} dt} \quad . \tag{6.22}$$

Equation 6.22 does still not consider the stitch density. Therefore, a comparison between the specific values of the stitch density as well as the deviation of the modified impregnation model to the experimental results is shown in Figure 96. At first, this figure shows the high accuracy of the modified impregnation model. More than 82 % of the measurements fit with a deviation smaller than 5 % to the impregnation model. However, the comparison between the deviation of the model to the experimental results and the stitch density does not show any clear relation. It does not indicate any trend. The statistical evaluation has also shown that the stitch density does not have any effect, especially in comparison to the other parameters, on the fibre impregnation. A higher accuracy is not able to achieve in the impregnation model. Therefore, the stitch density was not implemented into the analytical impregnation model.

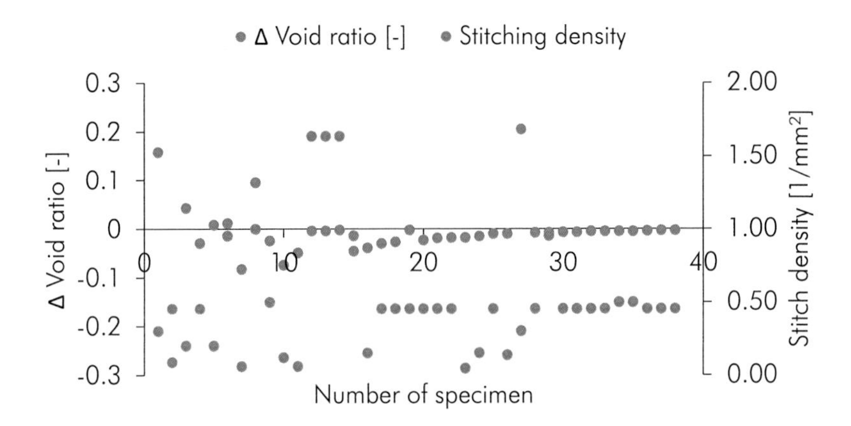

Figure 96. Deviation between the measured void content and the void content by the impregnation model with an empirical parameter related to the stitch density

In contrast to the stitch density, a great effect was observed for the process temperature. Whenever the temperature was chosen at 135 °C or smaller, a high degree of impregnation was achieved (The void content is lower than 1 %). The high significance of the temperature is shown in Figure 97.

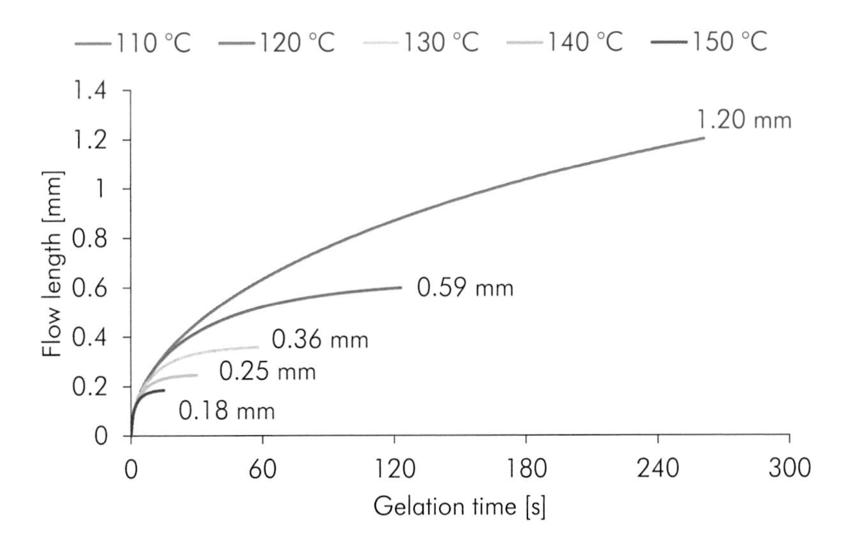

Figure 97. Development of the flow front with respect to the temperature

Figure 97 shows the development of the flow front with respect to the temperature by using a processing compression of 120 bar and a fibre volume content of 58 %. Lower temperatures lead to an extension of the flow front. In the first seconds, the lowest viscosity is available which leads to highest flow rate of the flow front. However, the increase of the viscosity due to cross-linking reactions leads to a reduction of the velocity and the flow front develops slower. The flow front ends at the specific gelation point. Figure 97 offers an illustration which thickness of the carbon fibre reinforcement can be completely impregnated by using specific temperatures. In addition, it gives information about the required cycle time.

Figure 98 shows the variation of the processing compression when different temperatures are used. The diagram shows processing compressions between 40 and 120 bar. Again, these curves consider a fibre volume content of 58 %. Figure 98 shows that the impact of varying the processing compression is smaller for higher temperatures which confirms the results of the statistical evaluation. The processing compression is not a highly significant parameter. The impact becomes higher when lower curing temperatures were used. The reason is the extension of the gelation point during the manufacturing of Hybrid SMC composites.

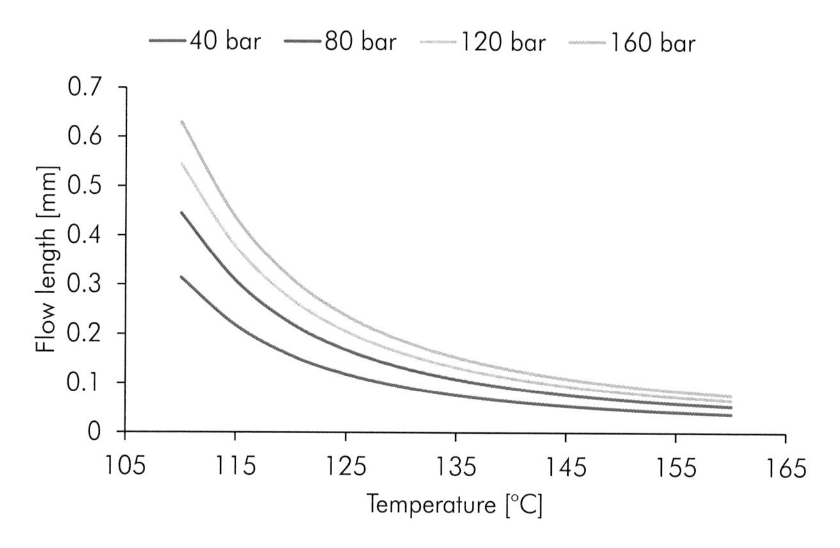

Figure 98. Flow front resulting by varying the processing compression and the temperature

All in all, the development of an impregnation model was presented. The impregnation model can be applied on the compression moulding of Hybrid SMC composites containing a non-impregnated carbon fibre reinforcement which is impregnated by the resin within the SMC material. The high accuracy of the model was demonstrated. It was shown how the flow length change with the variation of a parameter. The impregnation model is an important step for the acceptance of the new technology and it helps the operator to ensure the quality properties of the hybrid composite.

7 Demonstration of the Hybrid SMC construction

The Hybrid SMC way of construction is transferred to a demonstration part. This transfer illustrates the potential of using Hybrid SMC materials in future aircrafts or other industry sectors. A sidewall of the overhead stowage compartment was chosen to demonstrate the Hybrid SMC technology because the SMC material HUP63 has a material certification for the use in aircrafts' cabins. The development of the sidewall is described in the following. This chapter deals with the motivation, computation, the manufacturing, and the assembly of the overhead stowage compartment. A focus is set on the development of the carbon fibre reinforcements which are made by TFP.

7.1 The Overhead Stowage Compartment

Overhead stowage compartments (OHSC) have an important role in aircraft cabins. Especially on short-haul flights, passengers exclusively travel with hand luggage that is why recent research has focused on extending the space of OHSC. For example, in 2014, AIRBUS published a concept for overhead stowage compartments in the A320 which promises 60 % more space in comparison to the precursor [153]. Among other things, this was implemented by the use of the so called 'moveable bin' instead of the conventional 'fix bin'.
Despite the novel concept, the manufacturing methods are predominantly equal in comparison to the conventional methods. A sandwich way of construction is used in many areas of the aircraft cabin. The core is made out of a lightweight honeycomb sandwich which ensures sufficient stiffness within the part. The surface layers are made out of phenolic resin with reinforcing glass fibres. Phenolic resin is used because it has good thermal properties regarding heat distortion, heat conduction, and fire resistance which are mandatory properties for cabin materials (FST requirements). These parts are manufactured

either by autoclave processing or hand lay-up laminate. After the manufacturing of the sandwich panels, post-processing work has to be done. Then, the composite panels have to be manually folded and bonded. A further time-consuming step is the integration of the numerous metallic inserts. All in all, the manufacturing of the lightweight parts is dominated by many manual production steps which are responsible for the high production costs of aircraft interior parts. To compete especially with companies which are located in countries with lower production costs, new and innovative methods for the manufacturing of cabin elements have to be developed to fulfil the high aircraft requirements. Processing of SMC materials is a technology which enables new possibilities to design interior parts. For example, the FACC AG invest in a new production lane for the manufacturing of SMC parts for the entrance area of aircrafts [154]. If the use of SMC materials becomes establish in the future, the technology can be expanded to Hybrid SMC solutions. In the following chapter, the development of the Hybrid SMC way of construction is presented at a demonstration part using the example of the OHSC.

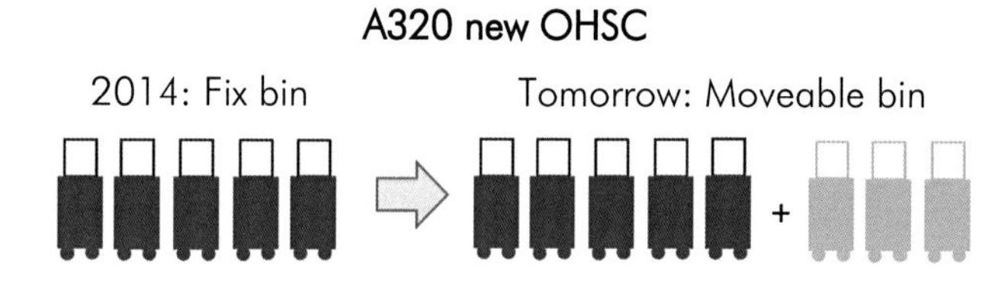

- 10 % more volume vs. Enhanced cabin in 2014
- Up to 60 % more bags

Figure 99. New concept for OHSC's published by AIRBUS in 2014 (According to: [153])

7.2 Computing and Manufacturing of the OHSC with a Hybrid Method of Construction

7.2.1 The bin type

Two main types of OHSC are available on the market today. One type is called 'fix bin' and has a fix housing with a moveable door. In contrast to the 'fix bin', the 'moveable bin' has a moveable housing with the integrated door. First, a decision has to be made which bin type should be used to develop the design with a hybrid way of construction. In the beginning of the part's development, it was decided to realize the 'moveable bin' because it has more complex mechanical loads at the areas of load introduction which can be perfectly absorbed due to an integration of TFP reinforcements. In addition, there are several load introducing areas at the sidewall of the OHSC. It was decided that only the sidewalls are developed in a Hybrid SMC way of construction. In contrast to the sidewalls, the back side of the OHSC is made out of the conventional honeycomb sandwich method. Prepreg materials cover the honeycomb sandwich. The sandwich material is cured in an autoclave. The assembly of back side with the Hybrid SMC sidewalls is realized by a developed groove and plug system.

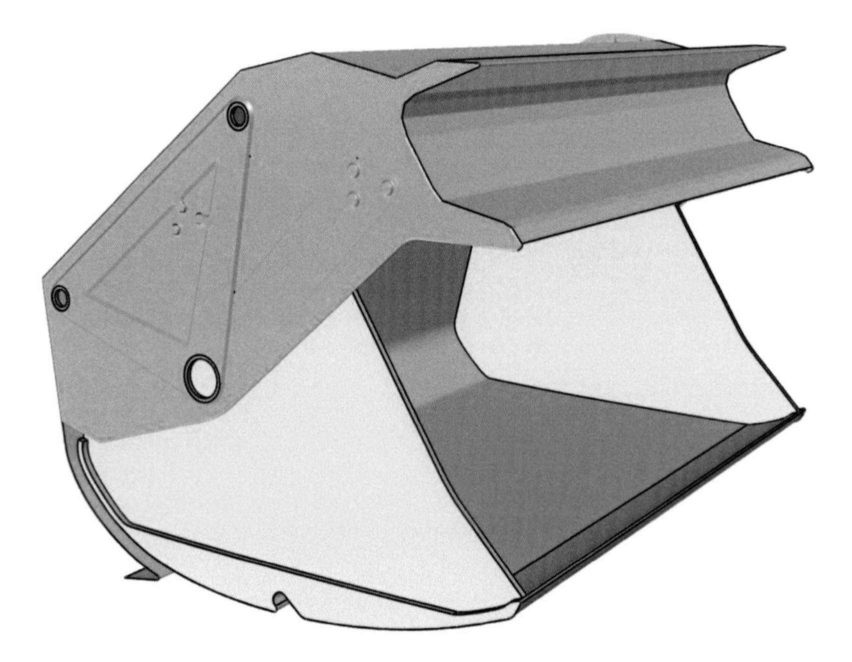

Figure 100. OHSC with sidewalls made by Hybrid SMC and conventional honeycomb back walls

7.2.2 The design of the TFP reinforcements

In the first step of the part design, the boundary conditions are defined. After the definition, design concepts are developed regarding the continuous carbon fibre reinforcement. Figure 101 shows the CAD model of the sidewall of the OHSC. With the help of the CAD model, a description is given for the different load introducing areas which are identified with the letters A – C.

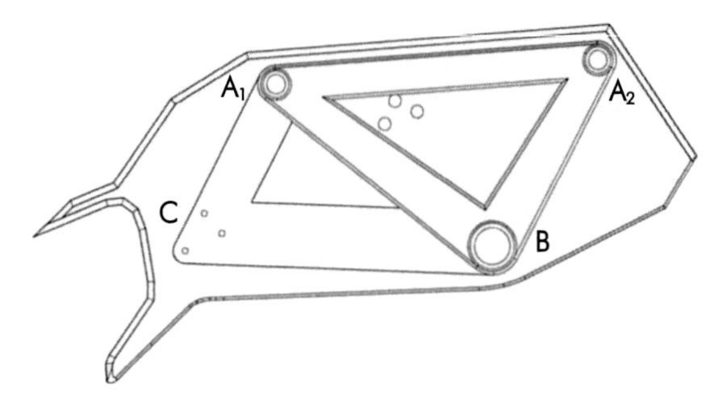

Figure 101. Sidewall of the OHSC (CAD model)

The sidewall of a moveable bin is characterized by two punctual load introducing areas with a diameter of 20.5 mm in the upper area of the sidewall (A). These areas are required to join the OHSC with bracket elements which fasten the OHSC to the structure. Another load introducing point is the lower area of the sidewall. This is the pivot point for the door (Figure 100) and has a diameter of 30.0 mm (B). Moreover, the sidewall contains six positions for metallic inserts which have smaller dimensions (d < 6.0 mm) (C). The outer contours were already defined by the end user and the maximum thickness was determined to 10.0 mm. In addition, another requirement defines the total weight. In comparison to the conventional sandwich part, the weight of the Hybrid SMC solution is not allowed to overrun. Based on the mechanical load cases of the OHSC, a first design of the TFP reinforcements was developed. It was decided that unidirectional carbon fibre loop structures are an appropriate method to handle the punctual tensile loads (A and B) [22]. In the beginning of the part development, the three load introducing points with the smaller diameter (C) were not strengthened with continuous carbon fibre textiles because of lower mechanical stresses. A simulation-based optimization was performed to improve this design and to quantify the use of the cost-intensive carbon fibres.

7.2.3 Computation

Finite Element Method (FEM) was used to optimize the detailed segmentation of TFP-structures and surrounding SMC materials. Taking damages by the needle as well as fibre waviness into account, the stitched continuous carbon fibre reinforcement was considered with reduced mechanical properties. A three-dimensional ply-model was used within the computation. The Hashin-failure criterion was used for the TFP structure which is impregnated by the resin of the SMC material [155]. As simplification, von Mises-stresses were considered for the pure SMC area due to the randomly oriented long fibre reinforcement which do not align during the manufacturing because a mould coverage higher than 90 % was used. A great advantage of the three-dimensional ply-modelling is the consideration of the stacking sequence and the adaption of the different materials in each ply. An optimization of the design was implemented by several design and computing iterations. The FEM has identified the critical regions within the hybrid structure (Figure 102) and finally indicated the specific area with its design and the amount of the continuous carbon fibre reinforcements [155].

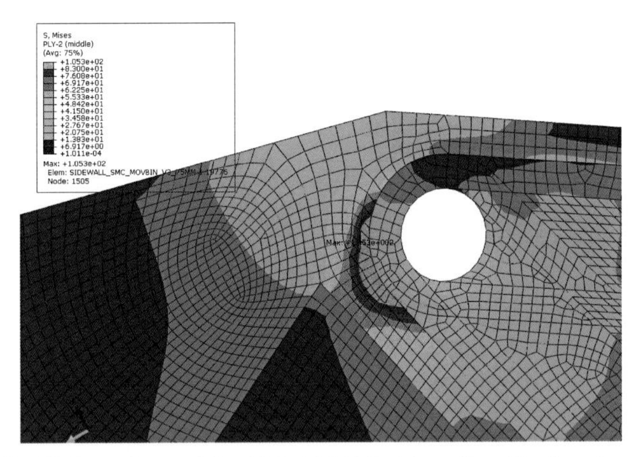

Figure 102. FEM analysis of the Hybrid SMC sidewall: critical regions [155]

One critical region was found in the interface between the carbon fibre reinforcements and the pure SMC material that is why a smooth transition between both materials has to be guaranteed. Furthermore, the simulations have shown that the area with the three smaller load introducing points would fail without any additional continuous carbon fibre reinforcement. Therefore, an area with additional unidirectional carbon fibre reinforcements was chosen to mechanically improve this area. The implementation of carbon fibre loops was not convenient in this case because of the small diameter at the load introducing points. A further reinforcement was integrated at the load introduction point 'A'. Despite the integration of carbon fibre loops, a premature failure was predicted by using the FEM. Therefore, additional unidirectionally arranged carbon fibre reinforcement was added at this area.

7.2.4 Manufacturing and Assembly

The manufacturing of the sidewall of the OHSC is divided into two production steps. First, the developed continuous carbon fibre reinforcement design is made by the TFP process. Then, the finished semi-finished textile is used within the compression moulding process with the semi-finished SMC at the same time to merge the materials to the Hybrid SMC part. In a first step, the final TFP design of the numerical simulation is implemented in an electronic stitching data. Due to the amount of the required continuous carbon fibre reinforcement and the final thickness of the TFP structure, which were evaluated by the FEM analysis, the reinforcements were separated into two parts. The usage of just one TFP textile would lead to a thick continuous carbon fibre reinforcement which would result in a long impregnation length. A complete fibre impregnation would be endangered. After the transfer of the developed structure to the TFP data, the continuous carbon fibre textile was manufactured. The first textile consisted of 12,540 stitches and the second one only consisted of 5,948. The design of the second TFP textile was characterized by the three entangled carbon fibre loops. An optimization of the amount of stitches was not implemented yet. Regarding the first and greater TFP reinforcement, the fibre course of the stitching data and the final continuous carbon fibre reinforcement are shown in Figure 103.

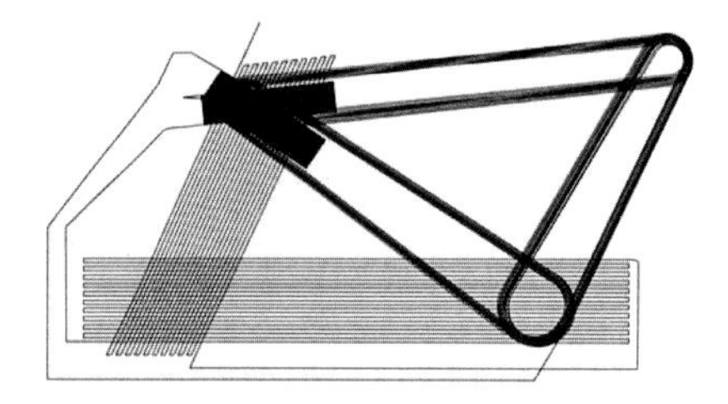

Figure 103. Fibre courses of the stitching data (left) and final implemented TFP textile (right, [156])

The following stacking sequence was chosen to manufacture the sidewall of the OHSC with a Hybrid SMC way of construction (Table 14).

Table 14. Stacking sequence of SMC material and TFP reinforcements in demonstration part

	Material	Code
1.	HUP63 – 2000 g/m^2	(C2)
2.	HUP63 – 2000 g/m^2	(C1)
3.	TFP reinforcement – small	(R2)
4.	HUP63 – 4000 g/m^2	(C1)
5.	TFP reinforcement – large	(R1)
6.	HUP63 – 2000 g/m^2	(C1)

The codes within Table 14 represents the design of the TFP textile and the design and the surface weight of the SMC material. Both TFP designs were already presented. R1 represents the larger carbon fibre reinforcement (12,540 stitches) and R2 represents the smaller TFP reinforcement (5,948 stitches). In accordance to the TFP identification, the SMC layer C1 is the larger SMC ply which covers 90 % of the mould in the beginning. The SMC layer C2 just covers the triangle of loops. Two different weight contents of SMC were used. Most of the time, a surface weight of 2000 g/m² is used. A surface weight of 4000 g/m² is used between the two TFP reinforcements to ensure the impregnation of the carbon fibre reinforcements on both sides. The total weight of all SMC plies have to be in common with the original component weight (CAD) less the TFP reinforcement. Cutting of the SMC layers were implemented by an ultrasonic cutter. The outer and inner contour of the TFP textiles were stamped which includes the load introductions points. The complete stack of the Hybrid SMC composite was prepared including all metallic elements. After the complete material stack was put into the preheated mould, the process was started with a temperature of 145 °C and a processing compression of 135 bar. The curing cycle was finished after five minutes. Before the assembly step was completed, deflashing was necessary after the demoulding of the part (Figure 104).

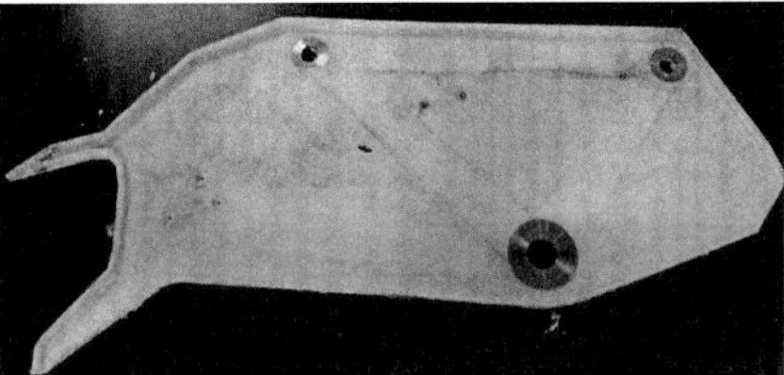

Figure 104. Mould with the semi-finished materials to process a Hybrid SMC part (left) and final part (right)

The back walls of the OHSC were made by the conventional honeycomb sandwich material with glass fibre prepreg surfaces. Due to a groove and plug assembly, the cured panels require an additional milling, folding, and bonding. The groove and plug system of the honeycomb sandwich panels and the Hybrid SMC sidewall is shown in Figure 105.

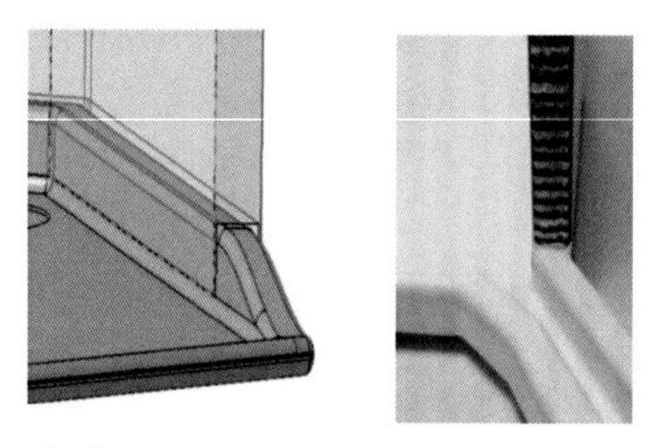

Figure 105. Groove and plug system within the OHSC

An additional layer of paint can be applied on the final part (Figure 106). All in all, the successful implementation of the Hybrid SMC way of construction was proven with the help of the OHSC demonstration. A transfer on other components can analogously performed.

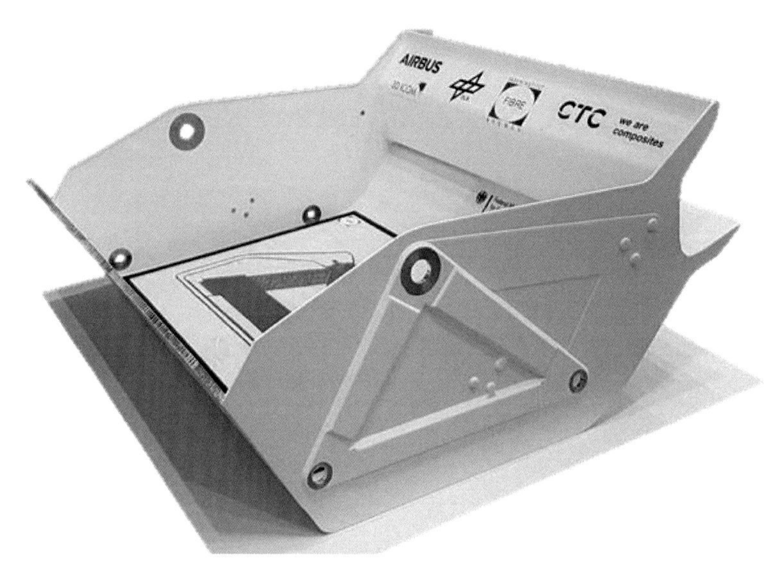

Figure 106. Final OHSC consisting of Hybrid SMC sidewall [156]

8 Conclusion and Outlook

In this study, an impregnation model was developed for the compression moulding of Hybrid SMC composites. The Hybrid SMC composite is characterized by a combination of conventional semi-finished SMC materials and continuous carbon fibre reinforcements which were manufactured by Tailored Fibre Placement (TFP). To ensure the economical process properties of the SMC compression moulding, the load-optimized TFP reinforcements have no pre-impregnation instead the continuous carbon fibre bundles were impregnated during the compression moulding by a resin transfer from the SMC materials which act as resin carriers. Due to the continuous carbon fibre reinforcement, the mechanical properties of the composite can be significantly increased.
The impregnation model is based on an approach of the fluid dynamics. The Eulerian and Lagrangian approach uses the material's derivative to describe a fluid's flow with respect to time and the circumstances. Taking an unsteady and uniform flow during the fibre impregnation into account, the material's derivative reduces to an objective function which just contains the velocity of the resin flow. The velocity of the flow can be described by Darcy's Law which describes the flow of a fluid through a porous medium. Darcy's Law is dependent on the material's viscosity, the processing compression, the permeability, and the thickness of the continuous carbon fibre reinforcement. Especially the viscosity of the SMC material requires several experimental tests. First, the determination of the gelation point is of great importance because after passing the gelation point no more fibre impregnation can be implemented. Non-isothermal viscosity measurements were performed to determine the gelation point which was found at a degree of cure of 2 %. The next step was the development of the cure kinetic of the SMC material with respect to the gelation point. The development of the cure kinetic is based on non-isothermal DSC measurements to

determine the specific heat flow signals. The heat flow signals were transferred to an expanded Prout-Tompkins-model to fit the empirical parameters. The result determines the final period between the start and the end of the fibre impregnation of each curing temperature. After that, isothermal viscosity measurements were taken for several temperature conditions to determine the change of the complex viscosity with regard to time. Another model, which uses empirical parameters, was developed and fitted to the time- and temperature-dependent viscosity. The experimental design, based on an experimental database, has validated the impregnation model. It was demonstrated that the curing temperature as well as the reinforcement's thickness are the most significant parameters on the fibre impregnation. The validation has shown that the developed impregnation model overestimates the real void content. Therefore, an empirical parameter was introduced to fit the model to the experiments. In the end, the developed impregnation model has a high accuracy. More than 82 % of the measured void contents fit to the impregnation model with a deviation less than 5 %.

Further scientific work can build on the achievements of this study. The approach, that was demonstrated in this study, does not make a difference between macro, meso or micro impregnation instead it considers a consistent flow front. Future work can concentrate on a differentiation of the impregnation levels. The experimental investigations on load introducing fibre architectures has shown the complexity when the fibre direction changes and additional metallic inserts are used. Therefore, a transfer of the impregnation model on curved fibre architectures is not implementable offhand because of the non-exclusive fibre impregnation through the thickness. If mould coverages smaller than almost 100 % are used, the SMC materials fills the mould due to compression moulding which essentially changes the conditions of the fibre impregnation. The mould coverage as well as the insertion time or the mould's velocity are important parameters within the Hybrid SMC compression moulding and require further investigation on the impregnation of continuous carbon fibre bundles. The transfer of the analytical consideration to a FEM-based solution would be a consequential step.

All in all, Hybrid SMC composites are a promising way to manufacture high performance composites with high production rates at the same time. The impregnation of the continuous carbon fibre bundles by SMC materials is an essential quality characteristic of the highly-innovative composite. The approach has the capability of fundamentally changing aircraft production.

Bibliography

[1] W. Frenz, Beitrag zur Messung der Produktivität und deren Vergleich auf der Grundlage technischer Mengengrößen. Springer Verlag, 2013.

[2] S. Mishra et al., "Studies on mechanical performance of biofibre/glass reinforced polyester hybrid composites," Compos. Sci. Technol., vol. 63, no. 10, pp. 1377–1385, 2003.

[3] J. Fleischer and J. Nieschlag, "Introduction to CFRP-metal hybrids for lightweight structures," Prod. Eng., vol. 12, no. 2, pp. 109–111, 2018.

[4] J. A. Tompkins, J. A. White, Y. A. Bozer, and J. M. A. Tanchoco, Facilities Planning, 4. Edition. John Wiley & Sons, 2010.

[5] D. G. Lee, H. S. Kim, J. W. Kim, and J. K. Kim, "Design and manufacture of an automotive hybrid aluminum/composite drive shaft," Compos. Struct., vol. 63, no. 1, pp. 87–99, 2004.

[6] J. Fleischer, S. F. Koch, and S. Coutandin, "Manufacturing of polygon fiber reinforced plastic profiles by rotational molding and intrinsic hybridization," Prod. Eng., vol. 9, no. 3, pp. 317–328, 2015.

[7] M. M. Davoodi, S. M. Sapuan, D. Ahmad, A. Aidy, A. Khalina, and M. Jonoobi, "Concept selection of car bumper beam with developed hybrid bio-composite material," Mater. Des., vol. 32, no. 10, pp. 4857–4865, 2011.

[8] A. Trauth and K. A. Weidenmann, "Continuous-discontinuous Sheet Moulding Compounds - Effect of hybridisation on mechanical material properties," Compos. Struct., vol. 202, pp. 1087–1098, 2018.

[9] Y. S. Song, J. R. Youn, and T. G. Gutowski, "Life cycle energy analysis of fiber-reinforced composites," Compos. Part A Appl. Sci. Manuf., vol. 40, no. 8, pp. 1257–1265, 2009.

[10] C. Dong, "Effects of Process-Induced Voids on the Properties of Fibre

Reinforced Composites," J. Mater. Sci. Technol., vol. 32, no. 7, pp. 597–604, 2016.
[11] W. V. Liebig, C. Viets, K. Schulte, and B. Fiedler, "Influence of voids on the compressive failure behaviour of fibre-reinforced composites," Compos. Sci. Technol., vol. 117, pp. 225–233, 2015.
[12] H. Zhu, B. Wu, D. Li, D. Zhang, and Y. Chen, "Influence of Voids on the Tensile Performance of Carbon/epoxy Fabric Laminates," J. Mater. Sci. Technol., vol. 27, no. 1, pp. 69–73, 2011.
[13] A. B. Strong, Fundamentals of Composites Manufacturing: Materials, Methods and Applications, 2nd ed. Society of Manufacturing Engineers, 2008.
[14] A. Aguggiaro, "Development of a supersport car aerodynamic component made of short fiber CFRP, characterized by high integration level," in Proceedings of SAMPE Europe, 2017.
[15] S. Moore, "SMC adopted for rear door frame of Toyota's new Prius PHV," Plastics Today, 2017. [Online]. Available: https://www.plasticstoday.com/automotive-and-mobility/smc-adopted-rear-door-frame-toyota-s-new-prius-phv/163725208056557.
[16] P. M. Research, "Global Market Study on SMC: Compact Devices and Lightweight Vehicles to Drive Sales," 2019.
[17] E. Witten, M. Sauer, and M. Kühnel, "Composites - Marktbericht 2017. Marktentwicklungen, Trends, Ausblicke und Herausforderungen," Compos. 2017, no. September, 2017.
[18] K. Xtra, "Bewährtes Material für neue Problemlösungen," pp. 20–21, 2013.
[19] "digitaltrends.com." [Online]. Available: https://icdn2.digitaltrends.com/image/digitaltrends/dt-geneva-lamborghini-huracan- performante-8-2-510x510-c-ar1.jpg. [Accessed: 04-Apr-2020].
[20] D. D. I. für N. E.V, "Verstärkte härtbare Formmassen – Spezifikation für Harzmatten (SMC) und faserverstärkte Pressmassen (BMC) –Teil 1: Bezeichnung; Deutsche Fassung EN 14598-1:2005," 2005.
[21] B. Van Voorn, H. H. G. Smit, R. J. Sinke, and B. De Klerk, "Natural fibre reinforced sheet moulding compound," Compos. - Part A Appl. Sci. Manuf., vol. 32, no. 9, pp. 1271–1279, 2001.
[22] H. Schürmann, Konstruieren mit Faser-Kunststoff-Verbunden, 2. ed. Springer Verlag Heidelberg, 2005.
[23] F. B. Alvey, "Study of the Reaction of Polyester Resins with Magnesium Oxide," J. Polym. Sci., vol. 9, pp. 2233–2245, 1971.
[24] E. Sancaktar and E. Walker, "Effects of calcium carbonate, talc, mica, and glass-fiber fillers on the ultrasonic weld strength of polypropylene,"

J. Appl. Polym. Sci., vol. 94, no. 5, pp. 1986–1998, 2004.
[25] C. Keckl, J. Kuppinger, and F. Henning, "Noch weniger Gewicht," in: Kunststoffe, pp. 30–32, 2014.
[26] K. Sawallisch, "Compounding of sheet molding compound," Polym. Plast. Technol. Eng., vol. 23, no. 1, pp. 1–36, 1984.
[27] S. Hörold, "Phosphorus flame retardants in thermoset resins," Polym. Degrad. Stab., vol. 64, no. 3, pp. 427–431, 1999.
[28] M. B. Chan-Park and F. J. McGarry, "Tough Low Profile Additives in Sheet Molding Compound," Polym. Compos., vol. 17, no. 4, pp. 537–547, 1996.
[29] C.-C. M. Ma, C.-T. Hsieh, H.-C. Kuan, T.-Y. Tsai, and S.-W. Yu, "Effects of Molecular Weight and Molecular Structure of Low Profile Additives on the Properties of Bulk Molding Compounds," Polym. Eng. Sci., vol. 43, no. 5, pp. 989–998, 2003.
[30] M. Neitzel, P. Mitschang, and U. Breuer, "Handbuch Verbundwerkstoffe," in Handbuch Verbundwerkstoffe, 2nd ed., L. Medina, J. Mack, and M. Christmann, Eds. Carl Hanser Verlag GmbH & Co. KG, 2014, p. 576.
[31] G. Abts, Kunststoff-Wissen für Einsteiger, 3. ed. 2016.
[32] F. Teodorescu, "Sheet Molding Compounds. Chopped Strands Reorientation models," 2018.
[33] A. Kraemer, S. Lin, D. Brabandt, T. Böhlke, and G. Lanza, "Quality control in the production process of SMC lightweight material," Procedia CIRP, vol. 17, pp. 772–777, 2014.
[34] E. L. Rodriguez, "On the Thickening Mechanism of Unsaturated Polyesters by Alkaline Earth Oxides and Hydroxides," J. Appl. Polym. Sci., vol. 40, pp. 1847–1855, 1990.
[35] M. Gruskiewicz and J. Collister, "Analysis of the thickening reaction of a sheet molding compound resin through the use of dynamic mechanical testing," Polym. Compos., vol. 3, no. 1, pp. 6–11, 1982.
[36] M. Feil and A. Becker, "Prozesstechnologie zur Reduzierung der Reifezeit von duroplastischen Formmassen (Lean SMC)," 2017.
[37] D. Judas, A. Fradet, and E. Marechal, "Mechanism of the thickening reaction of polyester resins: Study on models," J. Polym. Sci. Polym. Chem. Ed., vol. 22, no. 11, pp. 3309–3318, 1984.
[38] I. Vansco-Szmercsanyi and A. Szilagyi, "Coordination Polymers from Polycondensates and Metal Oxides. II. Effect of Water Molecules on the Reactions of Polyesters with MgO and ZnO," J. Polym. Sci. Polym. Chem. Ed., vol. 12, pp. 2155–2163, 1974.
[39] M. Gruskiewicz and J. Collister, "Analysis of the thickening reaction of a sheet molding compound resin through the use of dynamic mechanical

testing," Polym. Compos., vol. 3, no. 1, pp. 6–11, 1982.

[40] T. Sueck, "Neue UP-Harze für das SMC- und BMC- Verfahren," Dissertation at Technical University of Cottbus, 2008.

[41] B. Burns, B. M. Lynskey, and K. . Gandhi, "Research Projects in Reinforced Plastics 'Variability in Sheet Molding Compound,'" in RPG 4th Conference, 1976.

[42] K. Balkoteswara Rao and K. S. Gandhi, "Analysis of Products of the Thickening Reaction between Polyesters and Magnesium Oxide," J. Polym. Sci., vol. 23, pp. 2305–2317, 1985.

[43] G. Mennig and K. Stoeckhert, Mold-Making Handbook. Carl Hanser Verlag GmbH Co KG, 2013.

[44] M. M. Gauthier, Engineered Materials Handbook , Desk Edition. ASM International, 1995.

[45] A. D. Evans, C. C. Qian, T. A. Turner, L. T. Harper, and N. A. Warrior, "Flow characteristics of carbon fibre moulding compounds," Compos. Part A Appl. Sci. Manuf., vol. 90, pp. 1–12, 2016.

[46] K. Schladitz et al., "Non-destructive characterization of fiber orientation in reinforced SMC as input for simulation based design," Compos. Struct., vol. 160, pp. 195–203, 2017.

[47] H. T. Kau, "Determination of the orientation of short glass fibers in sheet molding compound (SMC)," Polym. Compos., vol. 8, no. 2, pp. 82–93, 1987.

[48] H. Büttemeyer and A. Miene, Abschlussbericht EFFKAB: Effiziente Kabinenarchitekturen. 2019.

[49] Y. Li et al., "Modeling and Simulation of Compression Molding Process for Sheet Molding Compound (SMC) of Chopped Carbon Fiber Composites," SAE Int. J. Mater. Manuf., vol. 10, no. 2, pp. 2017-01–0228, 2017.

[50] M. Hohberg, B. Fengler, N. Meyer, L. Kärger, and F. Henning, "Bauteilentwicklung mittels virtueller Prozesskette von SMC-Strukturen mit lokalen Verstärkungen," in Proceedings of: Erfa-Kreis (VDMA) Technologie und Prozesse, 2018.

[51] M. R. Barone and D. A. Caulk, "Kinematics of flow in sheet molding compounds," Polym. Compos., vol. 6, no. 2, pp. 105–109, 1985.

[52] E. Comte, D. Merhi, V. Michaud, and J. A. E. Månson, "Void Formation and Transport During SMC Manufacturing: Effect of the Glass Fiber Sizing," Polym. Compos., vol. 27, no. 3, pp. 289–298, 2006.

[53] D. Ferré Sentis, L. Orgéas, P. J. J. Dumont, S. Rolland du Roscoat, M. Sager, and P. Latil, "3D in situ observations of the compressibility and pore transport in Sheet Moulding Compounds during the early stages of compression moulding," Compos. Part A Appl. Sci. Manuf., vol. 92, pp.

51–61, 2017.
[54] T. S. Lundström and A. Holmgren, "Dissolution of voids during compression molding of SMC," J. Reinf. Plast. Compos., vol. 29, no. 12, pp. 1826–1837, 2010.
[55] E. Cybulski, Plastic Conversion Processes: A Concise and Applied Guide. CRC Press, 2009.
[56] AVK, Handbuch Faserverbundkunststoffe, 3. Edition. AVK – Industrievereinigung Verstärkte Kunststoffe e.V., 2010.
[57] J. M. P. Q. Delgado, Diffusion Foundations, vol. 14. Trans Tech Publications Ltd, 2017.
[58] L. Orgéas and P. J. J. Dumont, Sheet molding compounds. Wiley Online Library, 2012.
[59] O. Türk, Stoffliche Nutzung nachwachsender Rohstoffe, 1st Editio. Wiesbaden: Springer Fachmedien, 2014.
[60] V. Hopp, Grundlagen der chemischen Technologie, 4th Editio. John Wiley & Sons, 2008.
[61] K. Matyjaszewski and T. P. Davis, Handbook of Radical Polymerization. John Wiley & Sons, 2003.
[62] R. Toorkey, K. Rajanna, and P. Sai Prakash, "Curing of Unsaturated Polyester: Network Formation," J. Chem. Educ., vol. 73, no. 4, pp. 372–373, 1996.
[63] F. R. Jones, "Unsaturated Polyester Resins," Brydson's Plast. Mater. Eighth Ed., pp. 743–772, 2016.
[64] Wiley-VCH, Ullmann's Polymers and Plastics: Products and Processes. John Wiley & Sons, 2016.
[65] J. L. Thomason, M. A. Wug, G. Schipper, and H. G. L. T. Krikort, "Influence of fibre length and concentration on the properties of glass fibre-reinforced polypropylene: Part 3 . Strength and strain at failure," Compos. Part A, vol. 27A, pp. 1075–1084, 1996.
[66] M. Shirinbayan, J. Fitoussi, F. Meraghni, B. Surowiec, M. Laribi, and A. Tcharkhtchi, "Coupled effect of loading frequency and amplitude on the fatigue behavior of advanced sheet molding compound (A-SMC)," J. Reinf. Plast. Compos., vol. 36, no. 4, pp. 271–282, 2017.
[67] M. Shirinbayan, J. Fitoussi, F. Meraghni, B. Surowiec, M. Bocquet, and A. Tcharkhtchi, "High strain rate visco-damageable behavior of Advanced Sheet Molding Compound (A-SMC) under tension," Compos. Part B Eng., vol. 82, pp. 30–41, 2015.
[68] I. Baker and M. F. Ashby, Fifty Materials That Make the World. Springer International Publishing AG, 2018.
[69] H. Lee, M. Huh, J. Yoon, D. Lee, S. Kim, and S. Kang, "Fabrication of carbon fiber SMC composites with vinyl ester resin and effect of carbon

fiber content on mechanical properties," Carbon Lett., vol. 22, pp. 101–104, 2017.
[70] C. Nony-Davadie, L. Peltier, Y. Chemisky, B. Surowiec, and F. Meraghni, "Mechanical characterization of anisotropy on a carbon fiber sheet molding compound composite under quasi-static and fatigue loading," J. Compos. Mater., vol. 53, no. 11, pp. 1437–1457, 2019.
[71] M. Tiefenthaler, P. S. Stelzer, and C. N. Chung, "Characterization of the fracture mechanical behavior of C-SMC materials," Acta Polytech. CTU Proc., vol. 18, pp. 1–5, 2018.
[72] M. Cabrera-Rios and J. M. Castro, "An Economical Way of Using Carbon Fibers in Sheet Molding Compound Compression Molding for Automotive Applications," Polym. Compos., vol. 27, no. 6, p. 2006, 2006.
[73] J. Palmer, L. Savage, O. R. Ghita, and K. E. Evans, "Sheet moulding compound (SMC) from carbon fibre recyclate," Compos. Part A, vol. 41, no. 9, pp. 1232–1237, 2010.
[74] J. Palmer, O. Ghita, L. Savage, K. E. Evans, and N. Park, "New automotive composites based on glass," Proc. 17th Int. Conf. Compos. Mater., 2009.
[75] T. Rademacker, M. Fette, and G. Jüptner, "Nachhaltiger Einsatz von Carbonfasern dank CFK-Recycling," Light. Des., vol. 11, no. 5, pp. 12–19, 2018.
[76] T. Witte, "Vliesstoffe aus recycelten Kohlenstofffasern für den Einsatz in der Luftfahrt - Erfahrungen, Anwendungsbeispiele und Ausblick," in Proceedings of 33. Hofer Vliesstofftage, 2018.
[77] T. D. Hapuarachchi, G. Ren, M. Fan, P. J. Hogg, and T. Peijs, "Fire retardancy of natural fibre reinforced sheet moulding compound," Appl. Compos. Mater., vol. 14, no. 4, pp. 251–264, 2007.
[78] A. Asadi, F. Baaij, R. J. Moon, T. A. L. Harris, and K. Kalaitzidou, "Lightweight alternatives to glass fiber/epoxy sheet molding compound composites: A feasibility study," J. Compos. Mater., vol. 53, no. 14, pp. 1985–2000, 2019.
[79] M. O. Seydibeyoglu, A. K. Mohanty, and M. Misra, Fiber Technology for Fiber-Reinforced Composites. Woodhead Publishing, 2017.
[80] T. Böhlke et al., Continuous – Discontinuous Fiber-Reinforced Polymers, An Integrated Engineering Approach. Cincinnati: Hanser Publications, 2019.
[81] D. Bücheler, "Locally Continuous-fiber Reinforced Sheet Molding Compound," Dissertation at University of Karlsruhe, 2017.
[82] A. Trauth, P. Pinter, and K. A. Weidenmann, "Acoustic Emission Analysis during Bending Tests of Continuous and Discontinuous Fiber Reinforced

Polymers to Be Used in Hybrid Sheet Molding Compounds," Key Eng. Mater., vol. 742, pp. 644–651, 2017.

[83] M. Schemmann, J. Lang, A. Helfrich, T. Seelig, and T. Böhlke, "Cruciform Specimen Design for Biaxial Tensile Testing of SMC," J. Compos. Sci., vol. 2, no. 1, p. 12, 2018.

[84] M. Zaiß et al., "Use of Thermography and Ultrasound for the Quality Control of SMC Lightweight Material Reinforced by Carbon Fiber Tapes," Procedia CIRP, vol. 62, pp. 33–38, 2017.

[85] A. Trauth and K. A. Weidenmann, "Mechanical properties of continuously-discontinuously fiber reinforced hybrid sheet molding compounds," in Euro Hybrid Materials and Structures, 2016, pp. 57–62.

[86] A. Helfrich, S. Klotz, F. Zanger, and V. Schulze, "Machinability of Continuous-Discontinuous Long Fiber Reinforced Polymer Structures," Procedia CIRP, vol. 66. pp. 193–198, 2017.

[87] M. Hohberg, "Experimental investigation and process simulation of the compression molding process of Sheet Molding Compound (SMC) with local reinforcements," Dissertation at Karlsruher Institute of Technology (KIT), 2018.

[88] J. Wulfsberg, A. Herrmann, G. Ziegmann, G. Lonsdorfer, N. Stöß, and M. Fette, "Combination of carbon fibre sheet moulding compound and prepreg compression moulding in aerospace industry," Procedia Eng., vol. 81, pp. 1601–1607, 2014.

[89] F. Gortner, L. Medina, and P. Mitschang, "Influence of Textile Reinforcement on Bending Properties and Impact Strength of SMC-components," KMUTNB Int J Appl Sci Technol, vol. 8, no. 4, pp. 259–269, 2015.

[90] M. Fette, M. Hentschel, F. Köhler, J. Wulfsberg, and A. Herrmann, "Automated and Cost-efficient Production of Hybrid Sheet Moulding Compound Aircraft Components," Procedia Manuf., vol. 6, pp. 132–139, 2016.

[91] A. Silva-Caballero, P. Potluri, D. Jetavat, and R. Kennon, "Robotic dry fiber placement of 3D Preforms," in Proceedings of: 19th International Conference on Composite Materials, 2013.

[92] C. D. Nguyen, C. Krombholz, and H. Ucan, "Vergleich von Legetechnologien für die automatisierte GLARE-Bauteilfertigung," in Proceedings of: Deutscher Luft- und Raumfahrtkongress, 2018, pp. 1–9.

[93] Z. Stickmaschinen, "ZSK Stickmaschinen," 2019. [Online]. Available: https://zh-cn.facebook.com/ZSK.Stickmaschinen/posts/2547370465310015. [Accessed: 05-Jun-2020].

[94] P. Schiebel, "Entwicklung von Hybrid-Preforms für belastungsgerechte CFK-Strukturen mit thermoplastischer Matrix," Dissertation at University of Bremen - Faserinstitut Bremen e. V., 2018.
[95] P. Mattheij, K. Gliesche, and D. Feltin, "Tailored Fiber Placement - Mechanical Properties and Applications," J. Reinf. Plast. Compos., vol. 17, no. 9, pp. 774–786, 1998.
[96] K. Uhlig, A. Spickenhauer, L. Bittrich, and G. Heinrich, "Development of a highly stressed bladed rotor made of a CFRP using the Tailored Fiber Placement Technoloogy," J. Mech. Compos. Mater., vol. 49, no. 2, pp. 201–210, 2013.
[97] P. J. Crothers, K. Drechsler, D. Feltin, I. Herszberg, and M. Bannister, "The design and application of Tailored Fibre Placement," Proc. 11th ICCM Gold Coast Aust., 1997.
[98] H. Büttemeyer, A. Solbach, C. Emmelman, and A. S. Herrmann, "Development of a bracket element with a hybrid method of construction (CFRP/Ti)," in Proceedings of: Euro Hybrid Materials and Structures, 2016.
[99] N. K. Naik and S. Shembekar, "Elastic Behavior of Woven Fabric Composites : I - Lamina Analysis," J. Compos. Mater., vol. 26, no. 15, pp. 2196–2225, 1991.
[100] K. Uhlig, "Beitrag zur Anwendung der Tailored Fiber Placement Technologie am Beispiel von Rotoren aus kohlenstofffaserverstärktem Epoxidharz für den Einsatz in Turbomolekularpumpen," Dissertation at TU Dresden, 2017.
[101] K. Uhlig, M. Tosch, L. Bittrich, A. Leipprand, A. Spickenheuer, and G. Heinrich, "Evaluation of the geometrical influence of the stitching yarn on the stiffness and stress distribution in continuous carbon fiber reinforced plastics made by Tailored Fiber Placement using Finite Element Analysis," Proc. 20th Int. Conf. Compos. Mater., 2015.
[102] H. Büttemeyer, M. Fette, S. Drewes, and A. S. Herrmann, "Vertical Deformation of Continuous Carbon Fibre Textiles in Hybrid Sheet Moulding Compound Processing," in Proceedings of: SAMPE Europe, Southampton, 2018, pp. 1–8.
[103] M. Fette, "High efficient material and process combination for future aircraft applications based on advanced sheet molding compound technologies," 2016.
[104] A. Trauth, "Characterisation and Modelling of Continuous-Discontinuous Sheet Moulding Compound Composites for Structural Applications," Karlsruher Institute of Technology (KIT), 2018.
[105] M. Kästner, M. Obst, J. Brummund, K. Thielsch, and V. Ulbricht, "Inelastic material behavior of polymers - Experimental characterization,

formulation and implementation of a material model," Mech. Mater., vol. 52, pp. 40–57, 2012.
[106] D. Krause, T. Wille, A. Miene, H. Büttemeyer, and M. Fette, "Numerical Material Property Characterization of Long-Fiber-SMC Materials," Proc. AST 2019 Int. Work. Aircr. Syst. Technol., pp. 1–10, 2019.
[107] S. K. Chaturvedi, C. T. Sun, and R. L. Sierakowski, "Mechanical characterization of sheet molding compound composite," Polym. Compos., vol. 4, no. 3, pp. 167–171, 1983.
[108] S. Whitaker, "Flow in porous media I: A theoretical derivation of Darcy's law," Transp. Porous Media, vol. 1, no. 1, pp. 3–25, 1986.
[109] M. Duhovic, P. Kelly, D. May, T. Allen, and S. Bickerton, "Simulating Comporession-Induced Resin Transfer from a non-woven into a dry fiber structure," in Proceedings of: 22nd International Conference on Composite Materials (ICCM22), 2019, p. 12.
[110] S. Bickerton and M. Z. Abdullah, "Modeling and evaluation of the filling stage of injection/compression moulding," Compos. Sci. Technol., vol. 63, pp. 1359–1375, 2003.
[111] K. Masania, B. Bachmann, and C. Dransfeld, "The compression resin transfer moulding process for efficient composite manufacturing," Proceeding 19th Int. Conf. Compos. Mater., 2011.
[112] Y. A. Tajima and D. G. Crozier, "Chemorheology of an epoxy resin for pultrusion," Polym. Eng. Sci., vol. 28, no. 7, pp. 491–495, 1988.
[113] P. J. Halley and M. E. Mackay, "Chemorheology of Thermosets - An Overview," Polym. Eng. Sci., vol. 36, no. 5, pp. 593–60+, 1996.
[114] N. Rudolph and T. A. Osswald, Polymer Rheology - Fundamentals and Application. München: Carl Hanser Verlag, 2014.
[115] S. B. ROSS- MURPHY, "Rheological Characterisation of Gels," J. Texture Stud., vol. 26, no. 4, pp. 391–400, 1995.
[116] J. D. Fan and L. J. Lee, "Optimization of Polyester Sheet Molding Compound. Part II: Theoretical Modeling," Polym. Compos., vol. 7, no. 4, pp. 250–260, 1986.
[117] Y.-J. Huang, T.-J. Lu, and W. Hwu, "Curing of Unsaturated Polyester Resins- Effects of Pressure," Polym. Eng. Sci., vol. 33, no. 1, pp. 1–17, 1993.
[118] A. Yousefi and P. G. Lafleur, "Kinetic Studies of Thermoset Cure Reactions : A Review," Polym. Compos., vol. 18, no. 2, pp. 157–168, 1997.
[119] K. De la Caba, P. Guerrero, A. Eceiza, and I. Mondragon, "Kinetic and rheological studies of an unsaturated polyester cured with different catalyst amounts," Polymer (Guildf)., vol. 37, no. 2, pp. 275–280, 1996.

[120] J. M. Kenny and M. Opalicki, "Processing of short fibre / thermosetting matrix composites*," Compos. Part A, vol. 27, no. 3, pp. 229–240, 1996.
[121] N. Altmann and P. J. Halley, "The effects of fillers on the chemorheology of highly filled epoxy resins: I . Effects on cure transitions and kinetics," Polym. Int., vol. 119, no. May 2002, pp. 113–119, 2003.
[122] P. J. Halley, "A New Chemorheological Analysis of Highly Filled Thermosets Used in Integrated Circuit Packaging," Appl. Polym. Sci., vol. 64, no. 1, pp. 95–106, 1996.
[123] C. Brauner, Analysis of process-induced distortions and residual stresses of composite structures, Science Re. Berlin: Logos Verlag, 2013.
[124] J. L. Vilas, J. M. Laza, M. T. Garay, M. Rodriguez, and L. M. Leon, "Unsaturated Polyester Resins Cure: Kinetic , Rheologic , and Mechanical-Dynamical Analysis . I . Cure Kinetics by DSC and TSR," Appl. Polym. Sci., vol. 79, pp. 447–457, 2001.
[125] R. Hardis, J. L. P. Jessop, F. E. Peters, and M. R. Kessler, "Cure kinetics characterization and monitoring of an epoxy resin using DSC, Raman spectroscopy, and DEA," Compos. Part A Appl. Sci. Manuf., vol. 49, pp. 100–108, 2013.
[126] V. Massardier-Nageotte, F. Cara, A. Maazouz, and G. Seytre, "Prediction of the curing behavior for unsaturated polyester – styrene systems used for monitoring sheet moulding compounds (SMC) process," Compos. Sci. Technol., vol. 64, pp. 1855–1862, 2004.
[127] J. L. Martin, "Kinetic analysis of an asymmetrical DSC peak in the curing of an unsaturated polyester resin catalysed with MEKP and cobalt octoate," Polym., vol. 40, pp. 3451–3462, 1999.
[128] E. L. Rodriguez, "The Effect of Free Radical Initiators and Fillers on the Cure of Unsaturated Polyester Resins," Polym. Eng. Sci., vol. 31, no. 14, pp. 1022–1028, 1991.
[129] T. R. Cuadrado, J. Borrajo, and R. J. J. Williams, "On the Curing Kinetics of Unsaturated Polyesters with Styrene," vol. 28, pp. 485–499, 1983.
[130] M. E. Ryan, "Rheological and Heat-Transfer Considerations for the Processing of Reactive," Polym. Eng. Sci., vol. 24, no. 9, pp. 698–706, 1984.
[131] L. J. Lee, L. F. Marker, and R. M. Griffith, "The rheology and mold flow of polyester sheet molding compound," Polym. Compos., vol. 2, no. 4, pp. 209–218, 1981.
[132] Y. - S Yang and L. Suspene, "Curing of unsaturated polyester resins: Viscosity studies and simulations in pre- gel state," Polym. Eng. Sci., vol. 31, no. 5, pp. 321–332, 1991.
[133] J. M. Castro and C. Macosko, "Kinetics and Rheology of typical reaction

injection molding systems," Chem. Eng. Mater. Sci., pp. 434–438, 1980.

[134] J. M. Castro and C. W. Macosko, "Studies of Mold Filling and Curing in the Reaction Injection Molding Process," Am. Inst. Chem. Eng., vol. 28, no. 2, pp. 250–260, 1982.

[135] M. Hohberg, A. N. Hrymak, and F. Henning, "Process Simulation of Sheet Moulding Compound (SMC) using an extensional viscosity model," Proc. 14th Int. Conf. flow Process. Compos. Mater., pp. 1–6, 2018.

[136] L. M. Abrams and J. M. Castro, "Powder Coating of Sheet Molding Compound (SMC) Body Panels," Polym. C, vol. 22, no. 5, pp. 702–709, 2001.

[137] A. C. Liakopoulus, "Variation of the Permeability Tensor Ellipsoid in Homogeneous Anisotropic Soils," Water Resour. Res., vol. 1, no. 1, pp. 135–141, 1965.

[138] S. G. Advani and E. M. Sozer, Process Modeling in Composites Manufacturing, vol. 52, no. 1. 2002.

[139] M. V Bruschke and S. G. Advani, "Flow of generalized Newtonian fluids across a periodic array of cylinders," J. Rheol. (N. Y. N. Y)., vol. 37, no. 3, pp. 479–498, 1993.

[140] T. G. Gutowski, T. Morigaki, and Z. Cai, "The Consolidation of Laminate Composites," J. Compos. Mater., vol. 21, no. 2, pp. 172–188, 1987.

[141] B. R. Gebart, "Permeability of Unidirectional Reinforcements for RTM," J. Compos. Mater., vol. 26, no. 8, pp. 1100–1133, 1992.

[142] T. G. Gutowski, Z. Cai, S. Bauer, D. Bucher, J. Kingery, and S. Wineman, "Consolidation Experiments for Laminate Composites," J. Compos. Mater., vol. 21, pp. 650–669, 1987.

[143] H. S. Sas, E. B. Wurtzel, P. Simacek, and S. G. Advani, "Effect of relative ply orientation on the through-thickness permeability of unidirectional fabrics," Compos. Sci. Technol., vol. 96, pp. 116–121, 2014.

[144] S. Drapier, A. Pagot, A. Vautrin, and P. Henrat, "Influence of the stitching density on the transverse permeability of non-crimped new concept (NC2) multiaxial reinforcements: Measurements and predictions," Compos. Sci. Technol., vol. 62, no. 15, pp. 1979–1991, 2002.

[145] K. Siebertz, D. Van Bebber, and T. Hochkirchen, Statistische Versuchsplanung: Design of Experiments (DoE). Springer-Verlag, 2010.

[146] O. Focke et al., "Multiscale non-destructive investigations of aeronautic structures : from a single fiber to complex shaped fiber-reinforced composites," Proc. 8th Conf. Ind. Comput. Tomogr., pp. 1–7, 2018.

[147] E. Scheffler, Statistische Versuchsplanung und Versuchsauswertung. Stuttgart: Deutscher Verlag für Grundstoffindustrie, 1997.

[148] R. Schlittgen, Regressionsanalysen mit R. Walter de Gruyter, 2013.

[149] H. Büttemeyer, A. Gosler, and A. S. Herrmann, "Einfluss der Faservorspannung während der Aushärtung auf das quasi-statische Zugfestigkeitsverhalten von CFK-Schlaufenstrukturen," Proc. 15th Chemnitzer Textiltechniktagung, 2016.
[150] A. Lang, O. Focke, and A. S. Herrmann, "Mechanical behavior of loops with small diameters," Key Eng. Mater., vol. 742, pp. 374–380, 2017.
[151] A. Lang, L. Husemann, and A. S. Herrmann, "Influence Of Textile Process Parameter On Joint Strength For Integral CFRP-Aluminum Transition Structures," Procedia Mater. Sci., vol. 2, pp. 212–219, 2013.
[152] R. A. Granger, Fluid Mechanics. Courier Corporation, 2012.
[153] A. Noticias, "Airbus launches new pivoting overhead stowage on the A320 Family," no. 155, 2014.
[154] FACC-AG_press, "Neuer Airbus-Großauftrag für FACC," FACC AG Homepage, 2018. .
[155] H. Büttemeyer, A. Dimassi, M. Fette, M. Hentschel, and A. S. Herrmann, "Using Finite Element Method for the Sizing of Hybrid Sheet Molulding Compounds," Proc. SAMPE Eur., 2017.
[156] M. Fette, H. Büttemeyer, D. Krause, and G. Fick, "Development of Multi-Material Overhead Stowage Systems for Commercial Aircrafts by Using New Design and Production Methods," SAE Int. J. Adv. Curr. Pract. Mobil., vol. 2, no. 2, pp. 755–761, 2020.

In the dissertation the results from the supervision of the following students' works are included:

A. Gosler, (2016), University of Bremen, Betrachtung von Vorspannkräften bei CFK-Schlaufenstrukturen hinsichtlich der Auswirkungen auf die mechanische Leistungsfähigkeit.

F. Bittner, (2017), Hochschule Niederrhein University of Applied Sciences, Integration von TFP Preforms in SMC.

C. Wuttke, (2017), University of Bremen, Optimierung eines Prozesses zur Integration von
TFP Preforms in SMC.

X. Schüll, (2018), Hof University of Applied Sciences, Permeabilität von TFP-Kohlenstofffaserpreforms unter Berücksichtigung des Hybride-SMC-Prozesses.

D. Rudolf, (2018), University of Bremen, Untersuchung von endlosfaserverstärkten SMC-Platten im Bereich von metallischen Krafteinleitungselementen.

List of figures

List of tables

Publications

Parts of this dissertation have been published in the following publications and conference presentations:

Fette M, Büttemeyer H, Krause D, Fick G, Development of Multi-Material Overhead Stowage Systems for Commercial Aircrafts by Using New Design and Production Methods, SAE Int. J. Adv. & Curr. Prac. in Mobility 2(2) (2020) pp. 755-761, https://doi.org/10.4271/2019-01-1858.

Krause D, Wille T, Miene A, Büttemeyer H, Fette M, Numerical Material Property Characterization of Long-Fiber-SMC Materials, in: Proceedings of International Workshop on Aircraft System Technologies, Hamburg, Germany, 2019

Büttemeyer H, Fette M, Drewes S, Herrmann AS, Vertical Deformations of Continuous Carbon Fibre Textiles in Hybrid Sheet Moulding Compound Processing, in: Proceedings of SAMPE Europe Conference, Southampton, United Kingdom, 2018

Büttemeyer H, Dimassi A, Fette M, Hentschel M, Herrmann AS, Using Finite Element Method for the Sizing of Hybrid Sheet Moulding Compounds, in: Proceedings of SAMPE Europe Conference, Stuttgart, Germany, 2017

Fette M, Hentschel M, Santafe JG, Wille T, Büttemeyer H, Schiebel P, New Methods for Computing and Developing Hybrid Sheet Molding Compound Structures for Aviation Industry, in: Procedia CIRP 66 (2017) pp. 45 – 50, doi:10.1016/j.procir.2017.03.289

Büttemeyer H, Solbach A, Emmelmann C, Herrmann AS, Development of a Bracket Element with a hybrid method of construction (CFRP/Ti), in: Proceedings of Euro Hybrid, Kaiserslautern, Germany, 2016

Büttemeyer H, Gosler A, Herrmann AS, Einfluss der Faservorspannung während der Aushärtung auf das quasi-statische Zugfestigkeitsverhalten von CFK-Schlaufenstrukturen, in: Proceedings of 15th Chemnitzer Textiltechnik Tagung, Chemnitz, Germany, 2016

Büttemeyer H, Schiebel P, Solbach A, Emmelmann C, Herrmann AS, Performance of various designs of Hybrid Loop-Loaded CFRP-Titanium Straps, in: Proceedings of 20th International Conference on Composite Materials, Copenhagen, Denmark, 2015

Bisher erschienene Bände der Reihe

Science-Report aus dem Faserinstitut Bremen

ISSN 1611-3861

1	Thomas Körwien	Konfektionstechnisches Verfahren zur Herstellung von endkonturnahen textilen Vorformlingen zur Versteifung von Schalensegmenten ISBN 978-3-8325-0130-3 40.50 EUR
2	Paul Jörn	Entwicklung eines Produktionskonzeptes für rahmenförmige CFK-Strukturen im Flugzeugbau ISBN 978-3-8325-0477-9 40.50 EUR
3	Mircea Calomfirescu	Lamb Waves for Structural Health Monitoring in Viscoelastic Composite Materials ISBN 978-3-8325-1946-9 40.50 EUR
4	Mohammad Mokbul Hossain	Plasma Technology for Deposition and Surface Modification ISBN 978-3-8325-2074-8 37.00 EUR
5	Holger Purol	Entwicklung kontinuierlicher Preformverfahren zur Herstellung gekrümmter CFK-Versteifungsprofile ISBN 978-3-8325-2844-7 45.50 EUR
6	Pierre Zahlen	Beitrag zur kostengünstigen industriellen Fertigung von haupttragenden CFK-Großkomponenten der kommerziellen Luftfahrt mittels Kernverbundbauweise in Harzinfusionstechnologie ISBN 978-3-8325-3329-8 34.50 EUR
7	Konstantin Schubert	Beitrag zur Strukturzustandsüberwachung von faserverstärkten Kunststoffen mit Lamb-Wellen unter veränderlichen Umgebungsbedingungen ISBN 978-3-8325-3529-2 53.00 EUR
8	Christian Brauner	Analysis of process-induced distortions and residual stresses of composite structures ISBN 978-3-8325-3528-5 52.00 EUR
9	Tim Berend Block	Analysis of the mechanical response of impact loaded composite sandwich structures with focus on foam core shear failure ISBN 978-3-8325-3853-8 55.00 EUR

10	Hendrik Mainka	Lignin as an Alternative Precursor for a Sustainable and Cost-Effective Carbon Fiber for the Automotive Industry ISBN 978-3-8325-3972-6 43.00 EUR
11	André Stieglitz	Temperierung von RTM Werkzeugen zur Herstellung integraler Faserverbundstrukturen unter Berücksichtigung des Einsatzes verlorener Kerne ISBN 978-3-8325-4798-1 44.00 EUR
12	Patrick Schiebel	Entwicklung von Hybrid-Preforms für belastungsgerechte CFK-Strukturen mit thermoplastischer Matrix ISBN 978-3-8325-4826-1 45.50 EUR
13	Mohamed Adli Dimassi	Analysis of the mechanical performance of pin-reinforced sandwich structures ISBN 978-3-8325-5010-3 52.00 EUR
14	Maximilian Koerdt	Development and analysis of the processing of hybrid textiles into endless-fibre-reinforced thermoplastic composites ISBN 978-3-8325-5060-8 49.50 EUR
15	Holger Büttemeyer	Impregnation of stitched continuous carbon fibre textiles by Sheet Moulding Compounds ISBN 978-3-8325-5354-8 48.50 EUR

Alle erschienenen Bücher können unter der angegebenen ISBN-Nummer direkt online (http://www.logos-verlag.de) oder per Fax (030 - 42 85 10 92) beim Logos Verlag Berlin bestellt werden.